Andrew Crombie Ramsay

The Physical Geology and Geography of Great Britain

Six Lectures to Working Men

Andrew Crombie Ramsay

The Physical Geology and Geography of Great Britain
Six Lectures to Working Men

ISBN/EAN: 9783743402317

Manufactured in Europe, USA, Canada, Australia, Japa

Cover: Foto ©berggeist007 / pixelio.de

Manufactured and distributed by brebook publishing software (www.brebook.com)

Andrew Crombie Ramsay

The Physical Geology and Geography of Great Britain

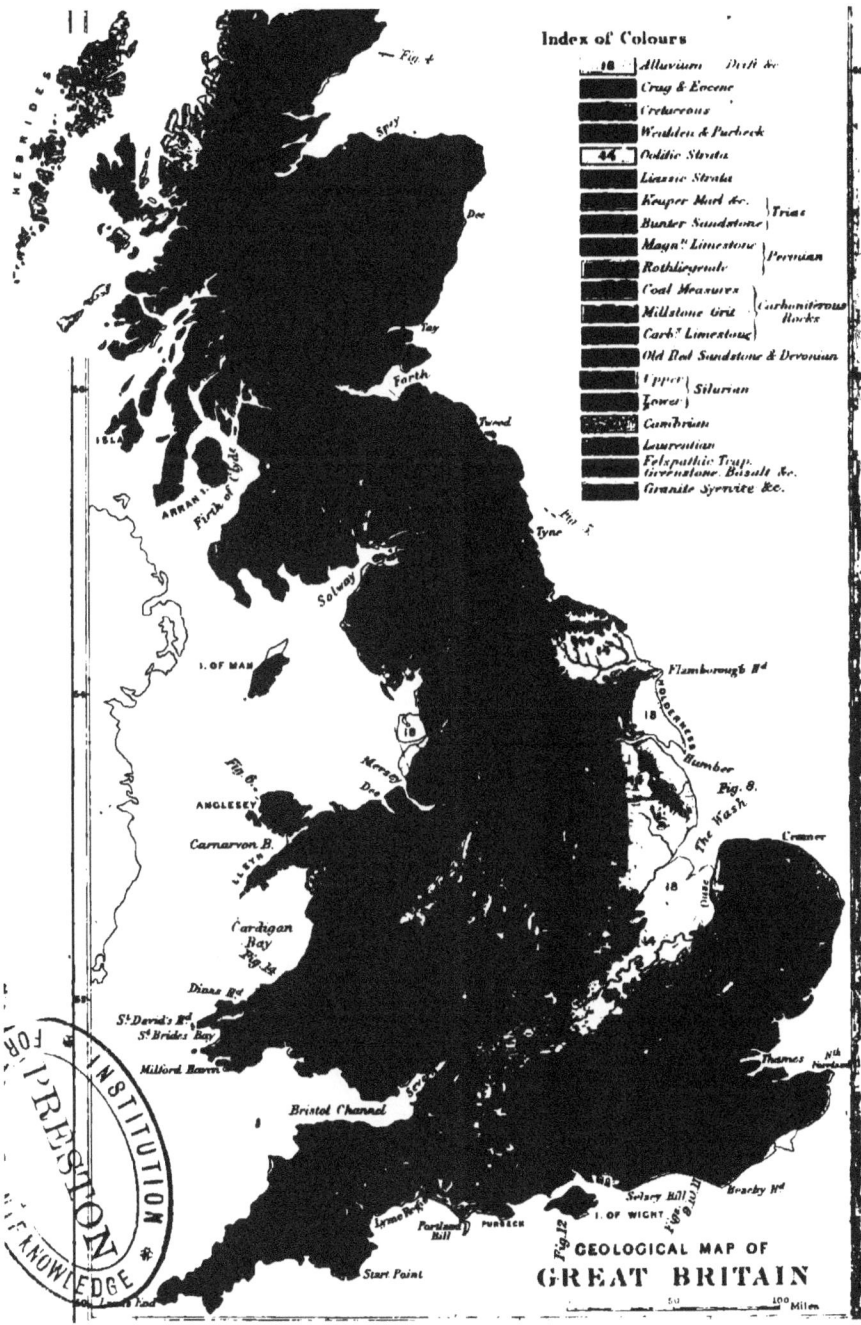

Index of Colours

GEOLOGICAL MAP OF
GREAT BRITAIN

THE

PHYSICAL GEOLOGY AND GEOGRAPHY

OF

GREAT BRITAIN.

SIX LECTURES

TO WORKING MEN

DELIVERED IN THE ROYAL SCHOOL OF MINES

In 1863.

BY A. C. RAMSAY, F.R.S.

LOCAL DIRECTOR OF THE GEOLOGICAL SURVEY OF GREAT BRITAIN.

SECOND EDITION.

LONDON:

EDWARD STANFORD, 6 CHARING CROSS.

1864.

There rolls the deep where grew the tree.
    O earth, what changes hast thou seen!
    There where the long street roars, hath been
The stillness of the central sea.

The hills are shadows, and they flow
    From form to form, and nothing stands;
    They melt like mist, the solid lands,
Like clouds they shape themselves and go.

TENNYSON.

LONDON
PRINTED BY SPOTTISWOODE AND CO.
NEW-STREET SQUARE

TO

THE MEMORY

OF

SIR HENRY THOMAS DE LA BECHE

C.B., F.R.S.

.

TO WHOSE EARLY TEACHINGS IN

PHYSICAL GEOLOGY

I AM SO MUCH INDEBTED,

THIS LITTLE BOOK

IS

AFFECTIONATELY DEDICATED

PREFACE.

THE first edition of these Lectures was printed from the shorthand report of Mr. I. Aldous Mays, and published with my consent. At his request I read and corrected the proof-sheets; but being much occupied at the time with other necessary work, and not in perfect health, many imperfections and mistakes, and a few positive errors, escaped my notice. In this edition the whole has been thoroughly revised, corrected, and in parts almost rewritten, and a good deal of fresh matter has been added, including a map, reduced for England from my own geological map, and for Scotland from the map by Sir Roderick Murchison and Mr. Geikie.

My object in delivering the original course, and in publishing this edition, has been to show how simple the geological structure of Great Britain is in its larger features, and how easily that structure may be explained to, and understood by, persons who are not practised geologists. Any one with a very moderate exertion of thought may thus realise the geological meaning of the physical geography of our country, and, almost without effort, add a new pleasure to those possessed before as he travels to and fro. The colours on geological maps will then no longer appear mysterious, but become easy to comprehend when associated with the geography of our island; and the little book may thus serve as a kind of condensed explanation of geological maps of Great Britain, and perhaps smooth the way for those who are just entering on the subject and feel alarm at its seeming difficulties.

ANDREW C. RAMSAY.

KENSINGTON: *January* 1864.

CONTENTS.

LECTURE I.

THE CLASSIFICATION OF ROCKS. DENUDATION.

In the good old days, those who thought upon the matter at all were perfectly content to accept the world as it is, believing that from its beginning to the present day it had always been much as we now find it, and that till the end of all things shall arrive, it will, with but slight modifications, always remain the same. But, by and by, when geology began to arrive at the dignity of a science, it was found that the world had passed through many changes; that the time was when the present mountains and plains were not, for the strata of which both are formed were once themselves sediments derived from the waste of yet older ranges. Thus it happens that that which is now land has often been sea, and frequently what is now sea has been land; and so there was a time before the existing rivers began to flow, and when all the lakes of the world, as we now know them, had no place on the earth. The whole subject is of the greatest interest,

B

and it is therefore my intention, in this course, to endeavour to show you—-taking our own island as an example—the reason why one part of a country consists of rugged mountains, and another part of low plains, or of high table-lands; why the rivers run in their present channels, and how the lakes that diversify the surface first came into being.

Experience tells me that at these courses of lectures a number of my old friends come to see me again and again, and also that there are many new faces present. Nevertheless, because so many of the old come to hear them, it is an object with me to vary the subjects as much as possible, so as to convey in each course some kind of instruction that was not given before. But as by the necessities of my position, and of my particular kind of knowledge, I am obliged to confine myself to subjects either purely geological, or intimately connected with geology, it is needful for the benefit of those who have not heard any of the previous lectures, that at the beginning I should enter on some of the rudimentary points of the subject, so as to make the remainder of the course intelligible to all. Therefore I begin to-night with an account of the origin of rocks; because it is impossible to understand the origin of the various kinds of scenery of our country, and to account for the classification of its mountains and plains without explaining the nature of the rocks which compose them.

Without further preface, then, all rocks are divided into two great classes—AQUEOUS and IGNEOUS—and there is a sub-class, which consists of aqueous rocks that have been altered, and which in their characters often approach some of those rocks that have been termed Igneous, in a popular sense, though in many respects very different from volcanic products. In this lecture I shall, however, confine myself to a general description of the two great classes of rocks; those of aqueous or watery origin, and those of igneous origin, which are the product of heat.

By far the larger proportion of the rocks of the world were formed by the agency of water. But, by what special processes were they formed? Every one knows that the rain which falls upon the land, draining the surface, first forms brooks, and that these brooks running into common channels and joining, by degrees become rivers; and every one who has looked at large rivers knows that they are rarely pure and clear,— notably, for instance, in the case of the Thames. Every river, in fact, carries sediment and impurities of various kinds in suspension or held in solution, and this matter, having been derived from the waste of the lands through which rivers flow, is carried to lower levels. Thus it happens that when rivers empty themselves into lakes, or—what is far more frequently the case— into the sea, the sediments that they hold in solution are deposited at the bottom of the lakes, or of the sea, as the case may be, and constantly increasing, they gradually form accumulations of more or less thickness, generally arranged in beds, or, as geologists usually term them, in strata. Thus, for instance, suppose any given river flowing into the sea. It carries sediment in suspension, and a layer will fall over a part of the sea-bottom, the coarser and heavier particles near the shore, while the finer and lighter matter will be carried out by the current and deposited further off. Then another layer of sediment may be deposited on the top of it, and another, and another, until, in the course of time, a vast accumulation of strata may in this manner be formed.

Again, if we examine the sea-coast where cliffs rise from the shore, we find that the disintegrating effect of the weather, and of the waves beating upon the cliffs gradually wears them away, comparatively quickly when made of clay or other soft strata, and in other cases very slowly perhaps, but still sensibly to the observant eye, so that in time, be they ever so hard, they get worn more and more backwards. The material derived from this waste when the cliffs are truly rocky, in the first instance, generally forms shingle at their base, as, for instance, with the pebbles of flint formed by the waste of the chalk. These being acted upon by the waves, are rolled incessantly backwards and forwards, as every one who has walked much by the sea must have noticed; for when a large wave breaks upon the shore, it carries forward the shingle, rolling the fragments one over the other, and in the same way they recede with the retreating wave with a rattling sound. This continued action has the effect of grinding angular fragments into rounded pebbles; and, in the course of time, large amounts of loose shingle are often thus formed. Such material when consolidated forms conglomerate. If, also, we examine with a lens the fragments that compose such a rock as sandstone, we shall find that it is formed of innumerable grains of quartz, and that these grains are often not angular but more or less rounded; and if you take up a handful of sea-sand and examine it in the same manner, you will frequently find that it does not consist of a quantity of small angular fragments, but of grains, the edges of which have been worn off by the action of the waves moving them constantly backwards and forwards upon themselves. Thus the little particles rubbing for ages upon each other, their angularity is gradually worn off, and they become grains, like rounded pebbles in shape, only much smaller. In this manner a very large amount of mechanical sediments are forming and have been formed.

If we examine the rocks that form the land, we very soon discover that a large proportion of them are

arranged in layers or bands of *shale, sandstone,* or
conglomerate, in a manner analogous to that which I
have just described as taking place at the mouths of
rivers and in the sea, thus proving that these layers
have been formed by the action of water. Take, for
instance, a possible cliff by the sea-shore, and we shall
probably find that it is made of a number of strata,
which may be horizontal, as in fig. 1, or inclined, or
even bent and contorted into every conceivable variety
of form, as in parts of figs. 3 and 4. If, as in the following diagram, we take a particular bed, No. 4, we

Fig. 1.

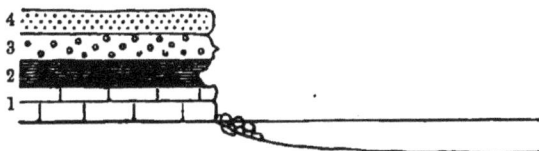

may find that it consists of sandstone, formed of a
number of differently-coloured layers arranged one
upon the top of another. Bed No. 3 may be of coarser
pebbly material, also arranged in layers, but not so
regularly as in No. 4, because the material is coarser.
No. 2 may consist of beds of thin shale of the finest
material, also arranged in layers, but the material being
much finer, each individual layer may be as thin as a
sheet of paper. Then in No. 1, the next and lowest
deposit, we may have a mass of limestone, arranged in
massive beds, the whole in the aggregate forming one

cliff. Rocks, more or less of these kinds, compose the bulk of the British islands; and remember that these were originally loose stratified sediments, piled on each other often to enormous thicknesses, and consolidated and hardened by pressure and chemical action. In some cases they have since been still further altered by heat and other agencies, but sometimes they are almost undisturbed except by mere upheaval, while in other cases the beds have been violently broken and contorted.

Then comes the question: Under what special conditions were given areas of these rocks formed? When we examine them in detail, we generally find that most of them contain, more or less, fossils of various kinds, —shells, corals, sea-urchins, the remains of plants and fishes, &c., and more rarely of the bones of terrestrial animals. For instance, in the bed of sandstone, No. 4 (fig. 1), we might find that there are remains of sea-shells; occasionally—but more rarely—similar bodies might occur in the conglomerate, No. 3; frequently they might lie between the thin layers of shale in No. 2; and it is equally common to find large quantities of shells, corals, sea-urchins, encrinites, and various other forms of life in such limestones as No. 1, which, in a number of cases, are wholly, or very nearly, composed of entire or broken shells and other marine organic remains.

Now, though strata of limestone have, in great part, been mechanically arranged, yet it comparatively rarely happens that quantities of unmixed calcareous sediment have been carried in a tangible form by rivers to the sea, or yet that it has been directly derived from the waste of sea-cliffs. When, therefore, it so happens that we get a mass of marine limestone consisting entirely of shells, which are the skeletons of marine creatures, the conclusion is forced upon us that, be the limestone ever so thick, it has been formed entirely by the growth and death of marine animals. In many a specimen, for instance from beds called the *Carboniferous limestone*, the naked eye tells us that it is formed perhaps entirely of rings of Encrinites, or stone-lilies as they are termed; and in many other cases where the limestone is homogeneous, the microscope reveals that it is made of exceedingly small particles of organic remains. It sometimes happens that such beds of limestone attain the enormous thickness of five hundred feet, or even of from one to four thousand feet in vertical thickness.

I will not tell you at present how we attain to the knowledge of the enormous thickness of these strata, because it would lead to a geological discussion which is, to a great extent, foreign to my present object; so that I must ask you to believe and take for granted, that the fact is so. But where does all the lime come from by which these animals make their skeletons? If you analyse the waters of the rivers that run through our own valleys, you will discover that most of them consist of hard water—that is to say, it is not pure like rain-water, but contains a quantity of various kinds of salts in a state of chemical solution, the most important of which is generally lime; for the rain-water that falls upon the surface of the land percolates the rocks, and rising again in springs, carries with it, if the rocks be at all calcareous, a quantity of lime in solution. The reason of this is, that all rain in descending through the air takes up a certain amount of carbonic acid—one of the constituents, accidental or otherwise—of the air; and this carbonic acid has the power of dissolving the lime which, more or less, enters into the composition of a large proportion of stratified rocks. In this way it happens that springs are often charged with lime, in the form of what chemists call a soluble bi-carbonate, which is carried into the rivers, and finding its way to the sea affords material to shell-fish and other marine animals, through their nutriment to make shells, bones, and tissues; and thus it happens, that by little and little lime is abstracted from the sea-water to form parts of animals, which dying, frequently produce, by their skeletons or shells, immense strata of nearly pure limestone.

But it often happens that along with shells there
are various other sediments found in the form of mud
or sand carried from the land into the sea; and in
this case, instead of pure limestone being formed, you
get impure limestone, or mixtures of shells with
common mud, sand, or pebbles, as the case may be.
In one case, for instance, we have a mass of rock
formed of consolidated mud, and the shells of oysters;
and by reason of the oyster-shells we obtain a large
percentage of lime in this specimen. In like manner
many other varieties of material may be intermingled,
as it were, almost at random. Sometimes strata consist
of lime and sand, or of lime, sand, and pebbles, or of
any two or all of these, mixed or alternating till they
become tens, or hundreds, or thousands of feet thick;
but when the limestone is pure and formed of organic
remains, its formation must have taken place in a sea,
or more rarely in fresh water, in which other sediments
at that time and in that locality were not being formed.

The other class of rocks, to which I have alluded,
are termed Igneous, and form a much smaller proportion
of the outer rocks of the entire world. Thus, to take
England and Wales as an example: in North Wales, in
Merionethshire, Carnarvonshire, and Anglesea, a considerable proportion, perhaps a tenth part, of the rocks
are formed of igneous masses. The whole of the rest
of Wales, down to Pembrokeshire, contains almost none

whatever. But for twenty miles eastward of St. David's Head, we have igneous rocks more or less distributed. The same comparatively small proportion of igneous rocks is found in parts of Scotland and Cumberland, and they also exist in Derbyshire, Devon, and Cornwall; whereas, if we examine all the midland, southern, and eastern parts of England, we shall find no igneous rocks whatever.

Now I have to explain how we are able to distinguish igneous from aqueous rocks; and, in a general way, I would say, that we can do so because most of them are *unstratified,* and have other external and internal structures different from those of aqueous deposits. To take examples: If we examine the rocks from any existing volcano, we find that the lavas poured out by it are frequently vesicular. This vesicular structure is due to gases and watery vapour in the melted mass, and these expanding, in their efforts to escape, blow out the melted rock and form a number of small vesicles or bubbles, just as yeast does in bread, and this peculiar vesicular structure is never found in the case of stratified rocks. Here then experience tells that any rock with this structure once formed part of a melted mass. I may know another specimen, which is crystalline, to be part of an old lava stream, because some one who obtained it, and on whose word I rely, told me that such was the fact, or

I have seen such cases, and know that this structure is characteristic of some volcanic rocks, arising from the circumstance that, in cooling, the substances of which the lava is composed crystallised in distinct minerals according to their chemical affinities. Another specimen may be from a rock which no man ever saw in a melted state; because it was fused, and cooled, and consolidated long before any human being looked upon creation. It belongs to a period called in Geology that of the Coal-measures; and when I examine its structure I find that it is nearly the same as in a specimen previously alluded to. It has been vesicular, but is not so any longer, because it happens that the original vesicles have been filled by infiltration of carbonate of lime. The mass has in fact been long under ground, and was infiltrated by water that, percolating through limestone rocks, carried lime in solution into the once empty vesicles. In these empty vesicles it has been deposited as carbonate of lime. But it frequently happens that the carbonate of lime, after such a rock has been exposed on the surface, is dissolved out by the carbonic acid held in rain-water, which again carries it away in solution as a bi-carbonate of lime, and then such a specimen again assumes a vesicular character analogous to that of some modern lavas. Therefore I should presume that this was an igneous rock. Again, we find that igneous rocks, in cooling, become crystal-

line—although they do not all do so. The melted
mass, in the first.instance, consists of a number of
substances mingled together; but as it cools, these
substances, under certain conditions, are apt to ar-
range themselves according to their chemical affinities,
and the result is the development of various minerals
in the rock — as, for instance, feldspar and augite.
On cooling, the constituents re-arranged themselves;
like drew to like, and the result was crystals of
feldspar, and crystals of augite. When I go abroad
and examine other igneous rocks, where no volcanic
action has occurred in the memory of man, or even for
an incalculable number of years before his existence, I
find, as in the case of the specimen from the Coal-measures, a structure similar to that which I observe in
certain modern lavas, and infer their igneous origin.

Again, if I take a specimen of another lava—from
a volcano not long extinct in the Island of Ascension—I
find that it is arranged in layers which in some degree
bear a resemblance to those which I have described as
layers of stratification; but if I compare it with the
slag which flows from iron furnaces, I find that they
are still more like that. Slag is in fact nothing but
artificial lava, being part of the silica and alumina of
the original iron ore and its flux of lime melted up
together. It frequently assumes a ribbon-like structure,
as any one must have observed, who has noticed slag as

it flows out of the furnace in a number of different
coloured bands, and this old lava from the Island of
Ascension presents the same wavy ribbon-like appearance. When I go to Wales and examine in the Silurian
region some of the oldest known lavas in the world,
I discover a similar structure—an arrangement in
slaggy-like layers; and therefore I infer that they were
ancient streams of lava.

Now, what would be the effect of a melted mass
of igneous rock coming in contact with stratified rocks,
such as some of these upon the table? The effect
would naturally be that, if the heat were sufficiently
strong, and if it were long enough applied, the stratified rock at the point of contact would undergo some
kind of alteration. If you place a mass of sandstone in
an iron furnace—or, better still, if you examine the
sandstone floor of an iron furnace where a perpetual
heat has been kept up for a long series of years—
where in fact the floor of the furnace has been in
contact with substances which are more or less of the
nature of melted lavas; this floor is found to be
changed. The sandstone is no longer comparatively
soft, as it was in its original state, but it has been
metamorphosed, or baked, and turned into a substance
which is known to geologists as 'quartz-rock;' the
colour is discharged, it has become white and hard, and
breaks with a splintery fracture. If again we submit

rocks composed originally of clay, like shales or slates, to intense heat, they assume the appearance of a kind of porcelain, and so completely is this recognised by geologists, that the term applied to rocks thus altered, is that it has been 'porcelainised,' or baked like potter's clay.

When I come to places among the hills where igneous or trap rocks rise through layers of sandstones, perhaps in a vertical manner, or where they send out branches hither and thither in among the beds, if I examine the strata at the point of contact with these, I find that the stratified rock has often altered its texture and structure, and changed its colour: and as you recede from the point of contact, it gradually becomes softer and softer until it passes into ordinary shale or sandstone. Experience has shown me that this is the effect of artificial heat, and also by actual observation I know that it has taken place in volcanic countries; and once having arrived at this point of experience, I have very little difficulty in other cases in determining whether or not I am in the presence of an igneous or a stratified rock, altered or unaltered as the case may be. And thus is it that geologists, by a process of analysis, are enabled to determine that the whole rock-masses of the outer world consists of two great classes—one class being *Igneous* and the other *Aqueous*.

The next point to be considered is—Are rocks of

different ages? This they evidently are, and the
diagram, fig. 1, will assist us to make it clear. There
the bed No. 1 must be the oldest, and the next,
No. 2, a little younger, because it was deposited upon
one already formed, and which therefore lies below,
and so on to 3 and 4—taking the strata in different
stages. But that is not enough to know. We are
anxious to understand what is the actual history of the
different stages which these rocks represent. Now, if
we had never found any fossil remains, we should lose
half the interest of this investigation, and our discovery,
that the rocks were of different ages, would have only
a minor value. Let us turn again to the diagram. We
find at the base a bed No. 1, say of limestone, composed of shells, the shells in the upper part of the bed
lie above those in the lower part, and therefore these
shells, or any other organic remains you please, in the
lower part of the bed, were dead and buried before
the once living shells which lie in the upper part came
into the area. Above the bed of shale, No. 2, there is
another stratum, No. 3, a conglomerate, and then comes
the bed of sandstone, No. 4; therefore the shells in the
bed of shale, No. 2, are of younger date than those in
the bed of limestone, No. 1; the shells in the conglomerate, No. 3, are newer than those in the shale, and
those in the sandstone, No. 4, are latest of all; and
each of these particular forms had lived and passed

away in succession before the sediment began to be formed in the bed above. All these beds, therefore, contain relics of ancient life of different dates, each bed being younger or older than the others according to the manner in which we read them.

But if we leave a petty cliff and examine the rocks on a larger scale, what do we find? Let us take, for instance, the middle of England—from the borders of South Staffordshire and Warwickshire to the neighbourhood of London; then we discover that the whole series is made of strata, formed more or less in the manner which I have described, in successive stages, the middle and upper parts of which added together are represented in the table at page 20, and in colours on the map. Thus, through Warwickshire and South Staffordshire, we have rocks formed of New red sandstone; the red sandstone dips to the east, and is overlaid by New red marl; the red marl dips also to the east, under beds of blue clay, limestone, and brown marl, called the Lias; these pass under a great succession of formations of limestones, clays, and sands, &c., which geologists have termed Oolites; these, in their turn, are overlaid by beds of sand, clay, and chalk, named the 'Cretaceous strata;' which again, in their turn, pass under the Tertiary clays and sands of the London Basin. All these pass fairly under each other in the order thus enumerated. Experience has proved this, for though there are occasional interruptions, some of the formations being absent in places, yet *the order of succession is never inverted,* except where, by what may be called geological accidents, in some parts of the world great disturbances have locally produced forcible inversions of some of the strata. The Oolites, for example, when little disturbed, never lie *under* the Lias, nor the Cretaceous rocks under the Oolites.

It is, therefore, not merely that the mere surface of the land is formed of various rocks, but the several formations dip or pass under each other in regular succession, being, in fact, vast beds placed much in the same way as a set of books, placed flat on each other, and then slightly tilted up at one end, may slope in one direction. As further proof of this succession, I may refer to the London basin, where we have strata round London, called the London Clay. Well sinkers frequently bore several hundreds of feet through this, and invariably they come to the chalk beneath; and so, if in some other places we bored through the chalk, we should come to Oolites, and if we bored through the Upper Oolites, we should come to the Middle and Lower Oolites, and so on through the Lias and other strata; and if we go further west, we find older more disturbed formations, cropping out in succession to the surface. Vertical sinking therefore often proves practically, what we know theoretically, viz. the underground continuity of strata one beneath the other, so that our island is formed of a series of beds of rock, some of many hundreds and some of several thousands of feet in thickness, arranged in succession, the lowest formation being of oldest and the uppermost of youngest date.

As we proceed from west to east, and examine minutely the various kinds of fossils found in those successive formations, we soon discover that *they are not the same in all,* and that most of them contain *marine* organic remains, which are in each formation of genera and species more or less distinct from those in the formation immediately above or immediately below. There are also a few freshwater deposits, and all of the fossil-bearing formations, whether of freshwater or of marine origin, contain the remains of animals that lived and died in the waters of the respective periods.

After a minute examination, therefore, of the structure of our island, the result is that geologists are able to recognise and place all the rocks in serial order, so as to show which were formed first and which were formed latest, and the following is the result of this tabulation, omitting minor details.

TABLE OF THE BRITISH FORMATIONS.

Recent.

Tertiary, or Cainozoic, and Quaternary.	UPPER.	Post Pliocene.	{ River gravels, raised beaches, bone caves, &c.
		Newer Pliocene.	{ Glacial drift, &c. Norwich Crag.
		Older Pliocene.	{ Red Crag. Coralline Crag.
	MIDDLE.	Miocene.	
	LOWER.	Upper Eocene.	{ Hempstead beds. Various formations, chiefly freshwater.
		Middle Eocene.	{ Bracklesham and Bagshot beds.
		Lower Eocene.	{ London Clay, Woolwich and Reading beds.
Secondary, or Mesozoic.	CRETACEOUS.	{ Chalk. Upper Greensand. Gault. Lower Greensand.	
	WEALDEN SERIES.	{ Wealden. Purbeck Beds.	
	OOLITIC SERIES,	{ Portland Oolite. Kimeridge Clay. Coral Rag. Oxford Clay. Cornbrash. Forest Marble. Bath Oolite. Stonesfield Slate. Inferior Oolite.	
	and LIAS.	{ Upper Lias Clay and sand. Marlstone (Middle Lias). Lower Lias Clay and Limestone.	
	TRIASSIC.	{ Upper. New red marl (Keuper). Lower. New red sandstone (Bunter).	
Primary, or Palæozoic.	PERMIAN.	Permian.	{ Magnesian limestone. Rothliegende.
	CARBONIFEROUS.	{ Coal-measures. Carboniferous limestone and shales.	
	OLD RED SANDSTONE, & DEVONIAN.	{ Upper. Lower.	
	SILURIAN.	{ Upper Silurian. Lower Silurian.	
	CAMBRIAN.		
	LAWRENTIAN.		

The Lawrentian rocks, which are the oldest known formation in the world, lie in Scotland in some of the Western Isles and the western parts of Sutherland, and consist of gneiss in a very far advanced stage of metamorphism.

The Cambrian rocks, which succeed them, contain a few obscure fossils, and the area occupied by this series is not large, being chiefly confined to small parts of Shropshire and Wales and the north-western part of Scotland. If we examine the Silurian rocks, which come next in succession, and which occupy for the most part Wales and Cumberland, we there find the relics of a number of peculiar forms of life, which, in the lower and upper divisions of the series, are vastly developed, both numerically and specifically; so with the succeeding age, the Devonian; so with the Carboniferous and Permian epochs; then through the Trias and Lias to the Oolitic epochs and their fossils; then, still higher in the scale of time, we arrive at the Cretaceous series, and so on into the Eocene beds and higher Tertiary strata, till at last we come to the present age. It is not, however, my business, in lectures bearing specially on physical geography, to give you a description of the various organic forms that have lived through these ages: that can only be done in a regular course of palæontological lectures.

Thus by an analysis of the order of deposition of the

rocks and their contents, geologists—led by the researches of the father of modern geology, William
Smith—are enabled to come to the important conclusion, that each formation was marked by its own
peculiar forms of life; that is to say, that each
formation was in its time a sea-bottom or a series of
sea-bottoms, in which peculiar kinds of life flourished,
which life for some reason in part or altogether disappeared, before a new period commenced, in which new
species inhabited the waters, which in their turn also
slowly died out; and so on in successive stages, from the
oldest epochs, through the whole of the formations, until
at last we come to the epoch in which we are now living.

It was necessary to explain this, because I shall have
frequent occasion to speak of the rocks by their names,
and to show their physical relations to each other in a
scenic point of view, these relations being connected
with phenomena dependent on their ages.

But before starting on this new subject, I must explain the meaning of a term which I shall have occasion to use very frequently, namely, *Denudation.*

'Denudation,' in the geological sense of the word,
means the stripping away of rocks from the surface,
so as to expose other rocks that lie beneath them.

Water running over the surface wears away the
ground over which it passes, and carries away detrital
matter, such as pebbles, sand, and mud, and if this goes

on long enough over large areas, there is no reason why
any amount of matter should not in time be removed.
For instance, we have a notable case in North America
of a very considerable result from denudation, now
being effected by the river Niagara, where, below the
Falls, the river has cut a deep channel through the rocks,
about seven miles in length. The proofs are perfect
that the Falls originally began at the great escarpment
which is at the lower end of what is now this long
gorge; that the river, falling over this ancient cliff, by
degrees wore for itself a channel backwards, about a
hundred and sixty feet deep, through strata that on
either side of the gorge form a great plateau.

I merely give this instance to show you what I mean
by denudation produced by running water. At one
time the channel did not exist. The river has cut it
out, and in doing so, strata formerly one hundred and
sixty feet beneath the surface have been exposed by
denudation. Probable but very uncertain calculations
show that to form this gorge a period at the least of
something like ten or twelve thousand years has been
employed.

Fig. 2.

Now, refer to fig. 2, and suppose that we have
different strata, 1, 2, 3, and 4, lying horizontally one

above the other, together forming a mass several hundreds of feet in thickness. Running water in the state of a brook or river by degrees wears away the rocks more in one place than another, so that the formations or strata 3, 2, and 1 *are successively cut into* and exposed at the surface, and a valley may in time be formed. This is the result of denudation.

In another way rain-water charged with carbonic acid, falling age after age on limestone rocks such as the chalk, not only wears away part of these rocks by ordinary denudation, but also dissolves the lime and carries it. off in solution, thus by waste of the upper beds bringing the lower strata to the surface. The evidence of the former existence of the wasted beds of chalk is witnessed, by prodigious numbers of unworn flints, scattered on the surface, these insoluble flints having once formed interrupted beds well apart from each other in the mass of the denuded chalk.

The constant atmospheric disintegration of cliffs, and the beating of the waves on the shore, is also another mode by which watery action denudes and cuts back rocks. Caverns, bays, and other indentations of the coast, needle-shaped rocks standing out in the sea from the main mass of the cliff, are all caused by the long-continued wasting power of the sea, which first helps to destroy the land and then spreads the ruins in new strata over its bottom, in time to be consolidated and again upheaved into land. Denudation by this process has always to a great amount been produced.

It requires a long process of geological education to enable any one thoroughly to realise the conception of the vast amount of old denudations; but when we consider that, *over and over again,* strata thousands and thousands of square miles in extent, and thousands of feet in thickness *have been formed by the denudation of older* rocks, equal in extent to the strata formed by their waste, we begin to get an idea of the greatness of this power. The mind is then more likely to realise the vast amount of matter that in times comparatively quite recent has been swept away from the surface of any country before it has assumed its present form. Without much forestalling the subject of a subsequent lecture, I may now state that a notable example on a grand scale may be seen in the coal-fields of South Wales and of the Forest of Dean. These two coal-fields were once united, but have been separated by the agency of vast and long extended denudations, which have swept away strata thousands of feet thick over a large area of Wales and the adjacent counties.

Observation and argument alike tell us that we need have no hesitation in applying this reasoning to other areas, and thus we come to the conclusion that the greater portion of the rocky masses of our island have been arranged and re-arranged, under a slow process of the denudation of old, and the reconstruction of newer strata, extending *over periods* that seem to our finite minds to stretch into infinity.

To explain in some detail the anatomical structure of our island, as dependent on the nature of its strata and the alterations and denudations they have undergone, will be the main object of the present course; and if you have been able to follow me in what I have already said, I am sure you will understand what I shall have to say in the remaining lectures.

LECTURE II.

THE PHYSICAL STRUCTURE OF SCOTLAND.

I HAVE now come to that part of the course in which it will be my duty to explain the connection between the geological phenomena of Britain and the nature of its scenery. In this lecture it is my intention to describe that district of Great Britain which is most mountainous, and to explain to you why it is that for the most part Scotland is so rugged. In another lecture I shall have to show you that there is a strong contrast between the physical features of Scotland and that of the middle and eastern parts of England, and to explain why the features of these two districts are essentially so distinct.

In last lecture I commenced by explaining that all rocks are divided into two great classes, namely, those of Aqueous and those of Igneous origin, and I showed you how aqueous rocks may be determined by the circumstance, that a great many of them contain relics

of marine and other life, in the shape of fossil shells,
fish-bones, and various other kinds of organic remains.
Also they are what is termed stratified, that is to
say, arranged in beds or layers one upon the other, and
the materials of which these beds are composed
generally show traces of having been acted upon by
water; being rounded and worn by the action of the
waves of the sea, or by the running waters of rivers.
The other great class of rocks, termed Igneous, are frequently crystalline, and from the effects which they
produce upon stratified rocks when they are in contact,
the latter are often altered. Then by comparing
igneous rocks of old date with those of modern volcanic
origin, we are able to decide with perfect truth that
rocks which were melted long before the human race
appeared upon the world, are yet of truly igneous
origin; and all the solid world above the surface of the
sea consists of these two great classes of rocks. But
there is a third division, which I called a *sub-class*,
known as *metamorphic rocks*; that is to say, stratified
rocks which have undergone a very serious kind of
alteration. All stratified rocks as they assume the solid
form become, indeed, to a certain extent altered; for
originally they were loose sediments of mud, sand,
gravel, or of lime, spread abroad sometimes in lakes,
but chiefly over the sea bottom, for fresh-water beds
form but a small part of the strata of the earth. But

when these were accumulated, bed upon bed, till thousands of feet were piled one upon the other, then, by intense and long continued pressure, which alone is sometimes sufficient to harden strata, and by chemical changes which take place in the interior of the strata themselves, by degrees they have become changed into . hard masses, consisting of shale, sandstone, conglomerate, or limestone, as the case may be. But these have not always remained in the same condition in which they were originally consolidated, for it has often happened that disturbances have taken place of a powerful kind, and the originally flat strata have been bent into every possible curve, in some cases for instance as shown in the following diagram.

Fig. 3.

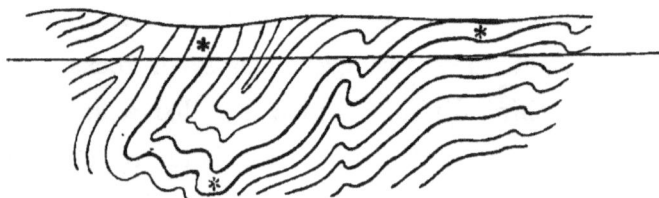

These are what is termed contorted strata when the disturbance has been extreme.

Now the metamorphic rocks, about which I have to speak, have been generally highly disturbed, and occupy a very large part of Scotland—I may say one-half—most of which includes, and lies north-west of the

Grampian mountains; and I must endeavour to explain by what processes metamorphism of rocks has taken place, not in detail, but simply in such a manner as to give you a general idea of the subject.

Metamorphic rocks, when the metamorphism is extreme, consist of gneiss and mica schist, chlorite schist, crystalline limestone, hornblende rock, and a number of others, which I need not name. It is enough for my present purpose if I make you understand that there are metamorphic rocks.

A typical specimen of gneiss consists of irregular laminæ of the minerals called *mica, quartz,* and *feldspar,* and it frequently happens that they are bent in a remarkable manner, or rather minutely folded in a great number of convolutions so small, that in a few feet of gneiss they may sometimes be counted by the hundred. Long ago all these rocks, that we term metamorphic, were, by the old geologists, called *Primitive* strata, and they were considered to have been formed in the earliest stages of the world's history, because in those countries that were first geologically described, they were found—or at least were supposed—to lie always at the base of the other strata, and from the peculiarity of the minute contortions in the gneissic rocks a theory now known to be erroneous was developed, which was this:

It is frequently found that granite and granitic rocks

are intimately associated with gneiss. Thus you will find, possibly, a mass of granite, with gneiss upon its flanks arranged in a number of small wavy folds or contortions. But granite is a crystalline rock, composed of feldspar, quartz, and mica, and the old theory was that the world at one time was in a state of perfect igneous fusion; but by and by, when it began to cool, the materials arranged themselves as distinct minerals, according to their different chemical affinities, and consolidated as granite. The great globe was thus composed entirely of granite, at all events at the surface; and by and by, as cooling still progressed, and water by condensation attempted to settle on the surface, that surface still remaining intensely heated, water could not lie upon it, for it was constantly being evaporated, and filled the atmosphere; but when the cooling became more decided, and consolidation had fairly been established, then water was able to settle on the surface of the mass of granite. But as yet it could not settle quietly like an ordinary sea of the present day; for by reason of the strong radiating heat, all the sea was kept in a boiling state, constantly playing upon the granite that rose above its surface here and there. The detritus thus worn from the granite by the waves of this primitive sea, was spread over its bottom; and because the sea was boiling, the sediment did not settle down in the form of regular layers, but became twisted and

contorted in the manner I have described.   All gneiss
therefore was conceived to be the original primitive
stratified rock of the world.

Subsequent research has shown that this theory will
not hold; for this among other reasons, that we now
know gneissic rocks of almost all ages in the geological
scale.   Thus in Scotland the gneissic rocks are of
Lawrentian and Silurian age; in Devon and Cornwall we have gneiss both of the Devonian and Carboniferous ages.   In the Andes, there are gneissic rocks of
the age of the chalk, and in the Alps, of the New red
Oolitic and Cretaceous series; and in 1862 I saw in
the Alps a species of gneiss of Eocene date, pierced by
granite veins, these strata being of the age of some of
the soft and often almost horizontal strata of the
London and Hampshire basins.   It is therefore now
perfectly well known to geologists that the term Primitive, as applied to gneiss, is no longer tenable ; for
we find rocks of every age metamorphosed, and therefore the old theory has been abandoned.

The oldest rock, however, in the British Islands is
gneiss, but that originally was doubtless a common
stratified rock of some kind or other.   In fact, as far
as the history told by the rocks themselves informs us,
we cannot get at their beginning at all, for all strata
are or have been made from the waste of rocks that
existed before; and this proves that the oldest stratified

rocks, whether metamorphosed or not, have a derivative origin. In some manuals the word Primary is still used as a convenient word to express the older strata, but no one now means by the term that they were the earliest rocks that ever existed, but simply that they are the *oldest rocks known.*

Now I must briefly endeavour to give you an idea of the theory of metamorphism. The simplest kind is of that nature which I spoke of in last lecture, namely, when an igneous is forced through or overflows a stratified rock, and remaining for a long time in a melted state, an alteration of the stratified rock in immediate contact with it takes place. Thus, sandstone may, by that process, become converted into quartz-rock, which is no longer hewable, like ordinary sandstone, but breaks with a hard and splintery fracture. It frequently happens also that when you find an igneous rock which passes through strata, the stratified rocks on each side for a certain distance, say a few inches, are altered, and imperfect crystals of some kind or other are developed where none existed before. Here then rocks are changed or metamorphosed a short distance from the agent that has been employed in effecting that metamorphism.

On a much larger scale, the sort of phenomena you meet with in a truly metamorphic region are as follows. In the midst of a tract of mica-schist, gneiss or other altered rocks, a boss of granite (or one of its allies)

D

rises, like that for instance of Dartmoor or of the north end of the Island of Arran. At a distance from the granite, the beds may consist, perhaps, of unaltered shale, or perhaps of slate, sandstone, and limestone. As you approach the granite, the limestones become crystalline, and often lose all traces of their fossils, the sandstones harden and pass into quartz-rocks, and the shale or slate loses its ordinary finely laminated texture, and passes by degrees into mica-schist, or gneiss, in which you find rudely alternating bands of quartz, feldspar, and mica, often arranged in gnarled or wavy layers. As you approach the granite still more closely, you find possibly that, in addition to the layers of mica, quartz, and feldspar, distinct crystals, such as garnet, staurolite, schorl, &c. are developed near the points of contact, both in the gneissic rock and in the granite itself.

It is not necessary for my argument that I should describe these minerals, beyond putting you in possession of the fact that such minerals are developed under these circumstances, and this is due to the influence of metamorphism.

Now, if we chemically analyse a series of specimens of clays, shales, slates, gneissic rocks and granites, it is remarkable how closely the quantities of their ultimate constituents will, in many cases, approach to each other. In all of them silica would form by far the largest proportion, say from 60 to 70 per cent.; alumina would come next, and then other substances, such as lime, soda, potash, iron, &c. would be found in smaller varying proportions; and what I now wish to impress upon you is, that the minerals developed in the gneiss, such as quartz, feldspar, mica, garnets, &c. *are not new substances* introduced into the rock, by contact with the granite, or by any other process; but *were all developed under the influence of metamorphism from materials that previously existed in the strata before the metamorphic action began.* Through some process, in which heat played a large part, the rock having been softened, and water—present in all rocks underground—having been diffused throughout the mass and heated, chemical action was set up, by which like drew to like, and the matter that composed the clay, shale, or slate, was enabled more or less to re-arrange itself, according to its chemical affinities, and minerals became visibly developed from elements that were in the original rock. This is a short sketch of the theory of metamorphism. I do not now attempt to give you the details, as that would occupy a lecture in itself.

Now to produce metamorphism, heat is necessary, to allow of internal movements by softening, without which I do not see how complete re-arrangement of matter and crystallisation could take place; and though it has always been easy to form theories about it, yet so little is known with precision about the interior of

the earth beyond a few thousand feet in depth, that
how to obtain the required heat is a difficulty. We
know, however, that strata which were originally deposited horizontally at the surface, have often descended thousands of feet towards the centre of the
earth, by gradual sinking, and the simultaneous piling
up of newer strata upon them. The layer that is
formed to-day is at the surface; but neither the land
nor the sea-bottom are steady; the land is in places
slowly descending beneath the sea, and the sea-bottoms
are themselves descending and shifting also. It has
frequently happened, therefore, that for a long period
a steady descent over a given area has taken place,
and simultaneously with this many thousands of feet
of strata have by degrees accumulated bed upon bed.
Every one knows that as we descend into the earth
the temperature increases, whence, in the main, the
theory of central heat has been derived. Heat increases
about 1° for every 60 feet, and the temperature therefore at so great a depth as 14,000 or 20,000 feet,
to which it could easily be shown some strata have
sunk, is much higher than at the surface. Furthermore, strata that were deposited horizontally have
been frequently disturbed and thrown into rapid contortions, or into great sweeping curves; and by this
means especially strata which once were at the surface
have been thrown, for aught we know to the contrary,

twenty, thirty, or forty thousand feet downwards, and therefore more within the influence of internal heat, as for instance in the bed marked* fig. 3. I do not wish you to understand that the globe is entirely filled with melted matter—that is a question still in doubt : but, were this a course of lectures on theoretical geology, I think I could prove that the heat in the interior of the globe sometimes in places apparently capriciously eats its way towards the surface by the fusion or alteration of parts of the earth's crust in a manner not immediately connected with the more superficial phenomena of volcanic action—and thus it may happen that strata which are contorted are in places brought within the direct influence of great internal heat. Under some such circumstances, you will easily understand how stratified rocks may have been so intensely heated, that they were actually softened; and all rocks being moist (because water that falls upon the surface percolates to unknown depths towards the interior of the earth), chemical actions were set going resulting in a re-arrangement of the substances which composed the sedimentary rock. Thus, common shale, or clay-slate, may have become changed into a mass of gneiss.

This theory of re-arrangement leads me to another question,—connected with, but not quite essential to my argument, as far as relates to physical geography,—

viz. what is the origin of granite, which in most manuals is classed as an igneous rock? For my own part, with some other geologists, I believe that in one sense it is an igneous rock,—that is to say, that it has been completely fused. But in another sense it is a metamorphic rock, partly because it is impossible in many cases to draw any definite line between gneiss and granite, for they pass into each other by insensible gradations; and on the largest scale, both in Canada and in the Alps, I have frequently seen gneissic rocks regularly interbedded with less altered strata, the gneiss being so crystalline that in a hand specimen it is impossible to distinguish it from some granitic rocks, and even on a large scale the uneducated eye will constantly mistake them for granites. Another very important circumstance is that *granite* and its allies *frequently occupy the spaces that ought to be filled with gneiss* or other rocks, were it not that the gneiss has been entirely fused. I therefore believe that granite and its allies are simply the result of the extreme of metamorphism brought about by great heat with presence of water.

One reason why it has been inferred that granite is not a common igneous rock, is, that *enveloping* the crystals of felspar and mica, there is generally a quantity of *free* silica, not crystallised out in definite forms like the two other minerals. Silica being far less easily fusible than felspar, it seems clear that had all the substances

that form granite been merely fused together in a dry
state, the silica ought on partial cooling to have crystallised first, whereas the felspar and mica have crystallised first, and the silica *not used in the formation of
these minerals* wraps them round in an amorphous
form. Therefore it is said that it was probably held in
partial solution in extremely hot water, even after
crystallisation by segregation of the other minerals
had begun. This theory, now held by several distinguished physical and chemical geologists, seems to me
to be sound, especially as it agrees exceedingly well
with the metamorphic theory as applied to gneiss—
granite being, as already stated, simply the result of the
extreme of metamorphism. In other words, when the
metamorphism has been so great that all traces of the
semi-crystalline laminated structure has disappeared, a
more perfect crystallisation has taken place, and the
result is a mass without any lamination in it.

Now in Scotland gneissic rocks and granites are
extensively developed. The north-western Highlands
and the Hebrides consist, to a great extent, of a formation which has been of late called Lawrentian,
named from a vast tract of gneissic rocks on the north
shore of the St. Lawrence, the geological age of which
was first determined by Sir William Logan. Above
them, in Scotland, other strata lie in the same district,
which, as they occur beneath the fossiliferous Silurian

series, are therefore supposed to be equivalent to the strata called Cambrian in Wales, and have received the same name. The Lower Silurian rocks come next in the series, and form nine-tenths of the Highlands of Scotland. They are chiefly gneiss and mica-schist, but have thick masses of quartz-rock at the base, inter-bedded with two bands of limestone, each of which con-tains fossils, and by this their age has been ascertained.

Next, on the north-east coast, we have the Old red sandstone, the Upper Silurian rocks which form such an important part of the English strata being absent.[*]

Above the Old red sandstone lie the Carboniferous rocks, consisting of Carboniferous limestone, and Coal-measures, the limestone forming in Scotland but a very small part of the series. These Carboniferous rocks lie in the great valley between the Grampian range on the north and the Lammermuir, Moorfoot, and Carrick hills in the south. Besides these formations there are others, in some of the Western Islands, such as Skye and Mull, and also on the east of Scotland and elsewhere. These consist of various members of the New red, Lias, and Oolitic strata, which, however, form such a very small part of Scotland, that they do not seriously affect its physical geography, and therefore I shall at present

---

[*] This order for the north of Scotland was first established by Sir R. Murchison. See 'Siluria' and Map of Scotland by Sir R. Murchison and Mr. Geikie.

say nothing about them, for I wish merely to put you in possession of the facts connected with the greater physical features of Scotland, omitting minor details.

Now, in the extreme north of Scotland, in Sutherland and Caithness, the manner in which these strata lie is shown in the following diagram. (See Map, line 4.)

Fig. 4.

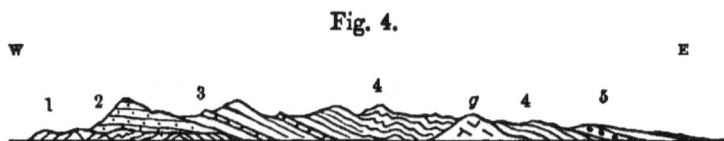

In some of the Western Isles from the Lewes to Bara, and in the north-west of Scotland from Cape Wrath to Gairloch, the country to a great extent consists of certain low tracts formed of Lawrentian gneiss (No. 1) twisted and contorted in the most violent manner. Upon this oldest gneiss the Cambrian rocks (2) lie, rising often into high mountains, which face the west in bold escarpments, and slope more gently towards the east. These strata frequently lie at low angles quite unconformably upon the older Lawrentian gneissic rocks; the meaning of this being, that the latter were disturbed, contorted, and extremely denuded before the deposition of the Cambrian strata upon them. The bottom beds of the latter consist of conglomerate of rounded pebbles, derived from the waste of this ancient Laurentian gneiss, which, therefore, is so old that it had been metamorphosed and was land before the deposition of the Cambrian strata. Upon these unaltered Cambrian beds, and again quite unconformably, the Lower Silurian strata are placed in the manner shown in the Diagram. The bottom beds consist of quartz-rock and two beds of limestone (3), the latter so altered that the fossils are sometimes with difficulty distinguishable, even by those most skilled in determining their nature. Then above the upper limestone we have a vast series of beds of mica-schist and gneissose rocks (4), mostly flaggy in the north-western region, but in the eastern parts of Sutherland, Aberdeenshire, &c. often so highly metamorphosed, that they are in many respects very similar to the more ancient Lawrentian gneiss.

Now these metamorphosed Silurian rocks, here and there associated with great bosses of granite and syenite (*g*) form by far the greater part of that extremely rocky region known as the *Highlands* of Scotland, stretching over brown heaths and barren mountain ranges, all the way from Loch Eribol on the north shore far south across the Grampians, to the Firth of Clyde on the west and Stonehaven on the east.

In Sutherland as a whole the Silurian strata dip eastward, and in Caithness we have the Old red sand-stone (5) lying quite unconformably upon the Si-lurian gneiss, and descending towards the sea. At its base the Old red sandstone consists of conglomerate,

not formed merely of small pebbles like
those of an ordinary shingle-beach, but frequently of huge masses sometimes yards in
diameter, mingled with others of smaller
size. All of them have evidently been derived from the partial destruction of those
ancient Silurian gneissic rocks (4) that
underlie the Old red sandstone.

Again, if you examine the Map of Scotland (line 5) you will find a broad band
of Old red sandstone running from Stonehaven on the east coast to Dumbarton on
the west, and there also masses of conglomerate lie at the base as in No. 2,
fig. 5. Overlooking this broad band, the
gneissose mountains No. 1 rise high into
the air; still reminding the beholder of
that ancient line of coast, against which
the waves of the Old red sandstone period
beat, and from its partial waste formed
that boulder-beach that now makes the
conglomerates. We are thus justified in
coming to the conclusion that the North
Highlands generally, formed land before
the time of the Old red sandstone, the
Grampian mountains as a special range
forming a long line running from north-east

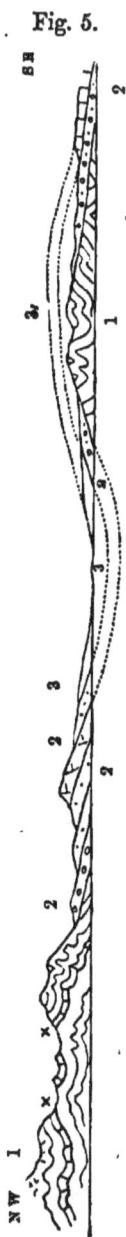

Fig. 5.

to south-west, the bases of its hills having been washed by the waters which deposited the Old red sandstone itself.

If you again examine the map you will find that a vast tract of country, forming half the *Lowlands*, stretches right across Scotland from north-east to south-west, including the Firths of Tay and Forth, and all the southern and eastern shores of the Firth of Clyde. This area is occupied by Old red sandstone and rocks of Carboniferous age (2 & 3, fig. 5), mostly stratified, but partly igneous. To the south lie the heathy and pastoral uplands known as the Carrick, Moorfoot, Pentland and Lammermuir hills, marked 1', which like the Highlands are also chiefly formed of Silurian rocks, but much less altered, and possessing nothing of a gneissic character. The Carboniferous rocks and Old red sandstone thus lie as a whole in a great hollow between the Grampian and the Lammermuir ranges, the coal-bearing strata consisting of alternations of shale, sandstone, ironstone, limestone, and coal, mingled with the volcanic products of the period.

Now how were the Carboniferous rocks formed? They consist of strata partly of fresh-water but chiefly of marine origin, for not only are the limestones formed chiefly of corals and shells, but many of the shales also yield similar fossils. Beds of coal are numerous, (whence the name Coal-measures, originally derived from the miners) and under each bed of coal there is a peculiar stratum, which often, but not always, is of the nature of fire-clay. Sometimes, it is called 'under-clay,' this being the miners' term, on account of its position beneath each bed of coal. Coal itself is well known to consist of mineralised vegetable matter, and when you examine the shales and sandstones associated with it, you frequently find in them quantities of vegetable remains, ferns, stems of reed-like plants (*Calamites*), trunks of various trees, &c. When the fire-clay is narrowly examined, you also generally find in it a number of portions of plants called *Stigmaria*, now known to be the roots of a fossil tree called *Sigillaria*, and this led Sir William Logan, Mr. Binney, and other geologists to infer that the under-clay was the original soil on which the plants grew, the decay and subsequent mineralisation of which formed beds of coal.

Among those plants which are found in the coal and its associated under-clay, and in the shale, may be enumerated the following genera: Sigillaria, Lepidodendron, Ulodendron, Calamites, Halonia, &c., and numerous genera of Ferns.

Now in the Scottish Coal-measures there are in Edinburghshire over 3,000 feet of coal-bearing strata, so that the lowest bed of coal may be nearly three thousand feet below the highest bed, in the centre of the basin, where the strata are thickest. Most of them

rise or 'crop,' as miners term it, to the surface some-
where or other, this 'out-crop' being the result of
disturbance of the strata and subsequent denudation,
and it is by means of this disturbance and denudation
that we are enabled to estimate the thickness of the
whole mass of strata, and to prove that one bed lies
several thousands of feet below another. Now, as the
'under-clay' contains roots, and was the soil on which
the plants grew, it is clear that the lowest bed of
coal was originally at the surface, and was formed by
the growth and decay of plants. After a time it seems
to have descended steadily and slowly, and other
strata were deposited upon it, sometimes in the sea, or
sometimes probably at the mouths of great rivers,
where a certain area was being filled with sediment.
By degrees a portion of the area, by filling up again,
became fit for the growth of terrestrial plants,
which plants decayed and formed another carbonaceous
stratum, that in its turn again sunk, and other strata
were deposited upon it. Vegetable growth again took
place, and so by intermittent sinkings and accumulations a great number of strata were produced, terrestrial, marine, estuarine and fresh-water, which by degrees
became a vast pile several thousands ' of feet thick.
The beds of vegetable remains were, probably, when first
formed, somewhat in the state of peat, and by immense
pressure and internal chemical changes, they in a long

lapse of time became mineralised, while by still later disturbance and denudation they are now in places exposed to view. In this way the Coal-measures were formed.

But in the Scottish area, during the formation of part of the Old red sandstone and of the Coal-measures, many volcanoes were at work; and thus, we have dykes and bosses of feldspathic trap and greenstone, and inter-stratifications of old lava streams, and beds of volcanic ashes mingled with the common sedimentary strata. These, being generally harder than the sandstones and shales with which they are interbedded, have more strongly resisted denudation, and now stand out in hilly ranges like the Pentland and Ochil Hills, or in craggy lines and bosses like Salisbury Crags, the Lomonds of Fife, and the Garlton Hills in Haddingtonshire, which give great diversity to the scenery, without ever rising to the dignity of mountains.

Having thus given a brief history of the mode of formation of the more important Scottish formations, you will already have begun to perceive what is the cause of the mountainous character of the *Highlands* and of the softer features of the *Lowlands.* It is briefly this : that in very ancient geological times, before the deposition of the Old red sandstone, the Silurian rocks which form almost entirely the northern half of Scotland, had already been metamorphosed and greatly disturbed. Such metamorphic rocks, though as a whole difficult of destruction, yet consist of inter-mingled masses of different degrees of hardness, whence the height of the mountains and their great variety of outline. In the south of Scotland from Galloway to the coast of Berwickshire, the same strata, forming the uplands of the Carrick, Moorfoot and Lammermuir hills, have been equally disturbed, but being comparatively unmetamorphosed, they are less hard, and have been more worn by denudation, whence their lower elevation. Then on the flanks of the Highland mountains, and partly round the eastern margin of what is now Scotland, the softer strata of the Old red sandstone, in various subformations, were deposited, formed partly, as the conglomerates testify, from the waste of the older Silurian strata. In time, the Old red sandstone period came to an end, and above that series—for it consists of several members, according to present nomenclature, which lie *uncon-formably* on each other—the Carboniferous rocks were formed. The whole were then again disturbed to-gether, a disturbance not confined to Scotland only, but embracing large European and other areas. In this lecture, however, I have merely to show you how these things affect the physical structure of Scotland.

But before the deposition of the Old red and Carbon-

iferous series, there is reason to believe that a wide
and deep valley already existed between the Grampian
mountains and the Carrick, Lammermuir, and Moorfoot
range; and in this hollow the Old red sandstone was
deposited, partly derived from the waste of the Silurian
hills on the north and south. But by-and-by, as deposition progressed, the land began to sink on the south,
and the upper strata of Old red sandstone overlapped
the lower beds, and began as it were to creep southwards across the Lammermuir hills, which, sinking still
further, were in turn invaded by the lower Coal-measures and Carboniferous limestone series. It appears,
therefore, from a consideration of all the circumstances
connected with the physical relations of the strata, that
the Coal-measures once spread right across the Lammermuir range, and were united to the Carboniferous
strata that now occupy the north of England; thus, with
part of the Old red sandstone, covering the Silurian
strata of the south of Scotland. This unconformable
covering has, however, in the course of repeated denudations, been removed from the greater part of that high
area, and now the Carboniferous strata are only found
in force in the great central valley through which flow
the rivers Forth and Clyde.

You will easily understand this if we refer to the section, fig. 5, across the central valley of Scotland from
the Grampian mountains to the Lammermuir hills.

E

The gneissic rocks (No. 1), with bands of Lime-stone marked *, of the Highlands, pass under the Old red sandstone (2), and rise again, highly disturbed, but not much metamorphosed, in the Lammermuir hills (1'). On these the Old red sandstone (No. 2) lies unconform-ably, above which come the Carboniferous rocks No. 3, lying in a wide broken and denuded synclinal curve. The diagram is, however, too small to show these breaks. The southern continuation of these strata once spread over the Lammermuir hills in a kind of anticlinal curve, in the manner shown by the dotted lines on the diagram below No. 3'.

Now why is it that the Carboniferous and Old red sandstone rocks have been specially preserved in the great valley, and almost entirely removed from the upland region of the Lammermuir hills? The reason is probably this.

When strata are thrown into a series of anticlinal and synclinal curves it frequently happens that those parts of the disturbed strata that are thrown downwards, so as to form deep basin-shaped hollows as in the bed or beds marked with an asterisk, * fig. 3, are by this means saved from the effects of denudation, while the upper parts of the neighbouring anticlinal curvatures have been denuded away.

In other words, in one place the beds lay so deep that,

being below the influence of the denuding agent, they
have escaped denudation, and the basin—as geologists
term it—remains; and this is the reason why so many
coal-fields lie in basins. It is not, as used to be supposed, that the beds were deposited in basins, but
that by disturbance, part of the strata have been thrown
into that form, and were saved from the effects of
denudation. Such basins are, therefore, equally common with all kinds of formations; though, because they
rarely contain substances of economic value, they have not
obtained the same attention from geologists that Coal-
basins have received. In the case now under review,
it happens that the Old red sandstone and Carboniferous
rocks lie in the hollow, while the continuation of part of
the same strata that lay high in an anticlinal form, and
originally spread over the Lammermuir hills (3′), has
been removed by denudation; the reason being that
during frequent oscillations of land, relatively to the
level of the sea, the higher ground was much more often
above water than the lower part. To understand this
thoroughly let us suppose, for the sake of argument,
that the whole of this country was underneath the
waters of the ocean, and then let it be raised to a
certain extent above the level of the sea. Part of the
Lammermuir area, *then covered with Coal-measures,*
rose above the water, and was immediately subjected to

the wear and tear of breakers on the shore, and of rain
and other atmospheric influences; while, on the other
hand, that portion that lay deep in the synclinal curve
was beneath the level of the sea, and thus escaped
denudation, because no wasting action takes place in
such situations.

By such geological *accidents* as these, the greater
features of Scottish scenery have been produced. The
Highlands are necessarily mountainous because they
are composed of disturbed rocks mostly crystalline;
which, having been often and long above water, have
been extremely denuded; such denudations having
commenced so long ago, that they date from before
the time of that extremely venerable formation, the
Old red sandstone, and have been repeated over and
over again down to the present day. Being formed
generally of materials of great but unequal hardness,
and associated with masses of granite, they have thus
been cut up into innumerable valleys, whence their
mountainous character; for mountains are rendered
rugged, less by disturbances of strata than by the
scooping out of the valleys. By *mere disturbance* of
strata, the land might rise high enough, but as our
mountain regions (and all others) now exist, it is by
a combination of disturbance of strata with extreme
denudation, that peaks, rough ridges, and all the cliffs
and valleys of the Highlands in their present form have

been called into existence. Farther south the different nature, both of the Silurian and newer rocks, coupled with other geological accidents, have produced the great valley, and the tamer but still hilly scenery of the Southern Lowlands.

LECTURE III.

THE geology of England and Wales is much more com-
prehensive than that of Scotland, in so far that it
contains a great many more formations, and its features,
therefore, are more various. England is the very
Paradise of geologists, for it may be said to be in itself
an epitome of the geology of almost the whole of
Europe. Very few known geological formations are
absent in England, and when they are so, with few
exceptions, these are of minor importance. In some
countries larger than England the whole surface is
occupied by one or two formations, but here, we
find all the formations shown in the column (page 20)
more or less developed. Those of Silurian age lie
chiefly in the north of England in Cumberland and
Westmoreland, and, in the west, in Wales and Cornwall.
Above them lie the Old red sandstone and Devonian
rocks, occupying vast tracts in Herefordshire, Worcestershire, South Wales, and in Devonshire and Cornwall.

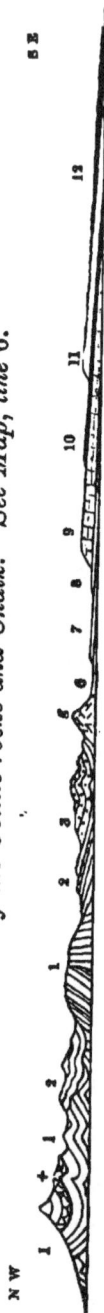

Fig. 6. Diagrammatic section from the Menai Straits across Wales, the Malvern Hills, and the Escarpments of the Oolitic rocks and Chalk. See Map, line 6.

Nos. 1 to 3 and *g* represent the disturbed Palæozoic mountainous country of North Wales, the adjacent counties on the east, and the Malvern Hills *g*.

6 to 8 the plains and slightly undulating grounds of the New red sandstone, red marl, and Lias.

9 and 10 the great Oolitic escarpment of the Cotswold hills, forming the first table-land.

11 the second great escarpment of the chalk, forming a second table-land, above which lie the Eocene strata, 12.

The upper Oolites close below the Chalk escarpment 11, are of less height relatively to the sea than the edge of the Oolitic escarpment at 9.

Above the Old red sandstone come the Carboniferous limestone and the Coal-measures, which in South Wales skirt the Bristol Channel, and stretch into the interior, while in the north they form a great backbone of country that reaches from the borders of Scotland down to North Staffordshire and Derbyshire. Other patches, here and there, rise from below the Secondary strata of the heart of England, and skirt the older formations in the west from Shropshire to Anglesey.

The general physical structure of our country from the coast of Wales to the Thames, will be easily understood by a reference to fig. 6 and to the following descriptions, and this structure is eminently typical, explaining, as it does, the physical geology of the chief part of England south of the Staffordshire and Derbyshire hills.

The Lower Silurian rocks of Wales (No. 1, fig. 6) and Cumberland consist chiefly of slaty and gritty strata, accompanied by, and interbedded with, various kinds of feldspathic lava and volcanic ashes, marked +, and mingled with them there are numerous bosses and dykes of greenstone, quartz-porphyry, and the like. These last, by their superior hardness, have helped to give that mountainous character to the western parts of our island, now called North Wales and Cumberland. In Pembrokeshire also, in a less degree, igneous rocks are largely intermingled with the Silurian strata, helping to form a very hilly country.

Without entering into details respecting the minor formations, known as the Lower and Upper Llandovery beds, it is sufficient to state that the Cambrian and Lower Silurian epoch ended in the British area by ·disturbance and contortion of the strata, and their up-heaval into land. This disturbance necessarily gave rise to long-continued denudations of this earliest English land, both by ordinary atmospheric agencies, and also by the action of the waves of the sea of a younger Silurian period, the evidence of which is seen in the conglomerates of the Upper Llandovery beds, which mingled with marine shells lie unconformably on the denuded edges of the Cambrian and Lower Silu-rian strata of the Longmynd, like an old consolidated sea beach. Slow submergence then took place beneath the Upper Silurian sea, in which the Upper Silurian rocks themselves were gradually accumulated uncon-formably on the Lower Silurian strata (2, fig. 6), till in places they attained a thickness of from three to six thousand feet.

Their uppermost strata then pass insensibly into the newer series known as the Old red sandstone (3, fig. 6), formed, if we include the entire formation, of beds of red marl, sandstone, and conglomerate, which, with a probable unconformable break in the middle, in turn again pass upwards in some regions into the Carbon-iferous limestone, which is overlaid in 'Wales and

England by the Millstone grit and the Coal-measures.*
This Carboniferous limestone is entirely formed of sea
shells, encrinites and other organic remains, and attains
a thickness of two thousand five hundred feet in South
Wales and the south-west of England, and in Derbyshire, where no man has ever seen its base, because
it rolls over in an anticlinal curve, it reaches even
a much greater thickness. The Millstone grit is in
South Wales 1,000 feet thick, and the true Coal-
measures, which are generally more or less of the
same nature as those that I described as occurring
in Scotland in the last lecture, are in Monmouthshire
and Glamorganshire not less than from 10,000 to
12,000 feet thick. The English Carboniferous rocks
differ from the Scottish beds in this, that in general
they have not been mixed with igneous interstratifications, except in Northumberland and Derbyshire,
where the Carboniferous limestone is interbedded with
ashes and lava, locally in Derbyshire called 'toad-stones.'
In South Staffordshire and in Coalbrook Dale, &c.,
there is a little basalt; but in Glamorganshire and
Monmouthshire, where the Coal-measures are thickest,
no igneous rock of any kind occurs. There and elsewhere in England the Coal-measures as usual consist

---

* This is not shown in fig. 6, but the Carboniferous limestone No. 4
is shown in fig. 7, lying, as it does in North Wales, unconformably on
Silurian rocks.

of alternations of sandstone, shale, coal, and ironstone; the coal everywhere being the remains of the decayed plants that grew upon the soils of the period, in the same way that I described them as growing in their day on what is now the Scottish Coal-measure area.

Next in the series come the Permian rocks (5, fig. 8), which however do not occupy so large a space in England as materially to affect the larger features of the scenery of the country. They form a narrow and very marked strip on the east of the Coal-measures from Northumberland to Nottinghamshire, where they frequently consist of certain beds of conglomerate and sandstone at the base, above which lies a long, low, flat-topped terrace of Magnesian limestone, the scarped edge of which faces west, and overlooks the lower undulations of the Coal-measure area. There are also certain other patches of Permian rocks known as the *Rothliegende,** here and there present on the borders of the North Wales and Shropshire coal-fields. The same formation, partly in the form of rough angular conglomerates, also lies on the Silurian rocks of the Malvern hills, and borders the coal-fields in the centre of England. But though these conglomerates here and there form

---

* A German name pretty generally adopted over Europe, for strata that were formerly called in England Lower new red sandstone. It is used to prevent confusion between these strata and the true New red sandstone, sometimes called the Bunter beds.

prominent points in the landscape, such as Wars Hill
on the Malvern range, and Frankly Beeches in South
Staffordshire, they produce no marked general feature
in the physical structure of the country, and therefore I
say little about them.

The Permian beds form the uppermost members
of that Palæozoic or old-life period, which begins in
England with the Cambrian rocks. The whole together
are sometimes called for the sake of convenience by the
old name of Primary strata. During the time they
were forming, this part of the world suffered many ups
and downs, accompanied by large denudations; and at
the close of the Permian period, a disturbance of the
strata on the greatest scale marks the end of this
great Palæozoic epoch over all Europe and more besides. By this disturbance, which was acccompanied
by much contortion of the strata, a large part of what
is now England was heaved up and formed dry land,
to be again wasted and worn away by sea-waves and
rivers, and all the common atmospheric agencies. This
old land in great part consisted of what we now know
as Wales, and the adjacent counties of Hereford, Mon-
mouth, and Shropshire, of part of Devon and Cornwall,
Cumberland, the Pennine chain and all the mountainous parts of Scotland. Around old Wales, and part
of Cumberland, and probably all round and over great
part of Devon and Cornwall, the New red sandstone

was deposited. Part at least of this oldest of the Secondary rocks was formed of the material of the older Palæozoic strata, that had then risen above the surface of the water, though it is not easy to trace precisely the whole of its subdivisions to the waste of special portions of the more ancient formations.

The New red sandstone series (No. 6, figs. 6 and 8) consists in its lower members of beds of red sandstone and conglomerate, more than a thousand feet in thickness, and above them are placed red and green marls, chiefly red, which in Germany go by the name of the Keuper strata, and in England are called the New red marl. The whole is often called the Trias. These formations fill the Vale of Clwyd in North Wales, and in the centre of England range from the mouth of the Mersey round the borders of Wales to the estuary of the Severn, eastwards into Warwickshire, and thence northwards into Yorkshire and Northumberland, along the eastern border of the Magnesian limestone. In the centre of England the unequal hardness of its subdivisions sometimes gives rise to minor escarpments, most of them looking west, over plains and undulating ground formed of softer strata. In the New red sandstone of Great Britain there are few relics of life, except at the very top where it passes into the Lias. They are plentiful in the Muschelkalk, which forms the middle part of the series in Germany, but is absent with us.

The Lias series (7 and 8, fig. 6), conformably suc-
ceeds the New red sandstone. The Lias constitutes
a well defined belt of strata, running continuously from
Lyme Regis on the south-west through the whole of
England, to Yorkshire on the north-east, and is an
extensive series of alternating beds of clay, shale, and
limestone, with occasional layers of jet. The Lias is
rich in the relics of ancient life, and it is in these strata
that those remarkable marine reptiles, the Ichthyosauri
and Plesiosauri, occur so plentifully. The unequal
hardness of the clays and limestones of the Liassic
strata causes some of its members to stand out in distinct minor escarpments, often facing west and north-
west. The Marlstone (8, fig. 6) forms the most prominent of these, and overlooks the broad meadows
of lower Lias clay that form much of the centre of
England.

Conformable to and resting upon the Lias are the
various members of the Oolitic series (9 and 10, fig. 6).
That portion termed the Inferior Oolite occupies the
base, being succeeded by the Great or Bath Oolite,
Cornbrash, Oxford clay, Coral Rag, Kimeridge clay and
Portland beds. These, and the underlying formations,
down to the base of the New red sandstone, constitute
what geologists term the Older Secondary formations,
and all of them, from their approximate conforma-
bility one to the other, occupy a set of belts of variable

breadth, extending from Devon and Dorsetshire northwards, through Somersetshire, Gloucestershire, and
Leicestershire, on to the north of Yorkshire, where
they disappear beneath the German Ocean.

When the Portland beds had been deposited (forming
the top of 10, fig. 6), the entire Oolitic series in what
is now the south and centre of England, and much
more besides in other regions, was raised above the
sea-level and became land; and because of this elevation, in the Isles of Purbeck, Portland and the Isle
of Wight, and in the district known as the Weald,
there is evidence of a state of affairs common in all
times of the world's history, but from causes that it
would take long to enumerate, very unusual as far as
preserved strata are concerned. In fact, we have here
a series of beds, consisting chiefly of clays, sands, sandstones, and shelly limestone, indicating by their fossils
that they were accumulated in an estuary where fresh-
water and occasionally brackish-water and marine
conditions prevailed. The position of these beds with
respect to the Cretaceous strata, you will find in fig. 10,
p. 78, marked *w, h,* and to prove that they are intermediate in date to the Oolites and Cretaceous rocks,
I may mention that in the Isle of Purbeck, they are
seen lying between the two. The Wealden and Purbeck beds, indeed, represent the delta of an immense
river, which in size may have rivalled the Ganges or

the Mississippi, and whose waters carried down to its mouth land-plants, small mammalia, and great terrestrial reptiles, and mingled them with the remains of shells, fish, crocodiles, and other forms native to its waters. I do not by any means wish you to understand that this immense river was formed simply by the drainage of the small territory we now call Great Britain. I do not indeed quite know where the mass of the continent lay through which it ran and which it drained, but I do know that England formed a part of it, and that in size it—that continent—must have been far larger than Europe and probably as large as Asia, or the great continents of North or South America.

I must, however, explain how we know that the Wealden series were accumulated under fresh-water conditions, and as a river deposit. The proof lies partly in the nature of the strata, but chiefly in the nature of the organic remains contained in them. The fish give no positive proof, but a number of Crocodilian reptiles give more conclusive evidence, together with the shells, most of them being of fresh-water origin, such as Unio, Paludina, Planorbis, Limnæa, Physa, and such like, which you may find living in many a river, pond, or canal of the present day. Some of these are so very like existing species that it requires all the skill of the accomplished palæontologist to tell that there is any difference between them. But now and then we find bands of marine remains, not confusedly mixed with the fresh-water deposits, but interstratified with them ; showing that at times the mouth and delta of the river had sunk a little, and that it had been invaded by the sea, so that oysters and other *salt-water* mollusca lived and died there. Then by gradual changes it was again lifted up, and became an extensive fresh-water area. It is important to mention these circumstances, because the nature of some of these half consolidated strata exercises a considerable effect on the amount and nature of the denudation of the rocks in the south-east of England, and consequently upon its scenery.

This episode at last came to an end, by the complete submergence of the Wealden area; and upon these fresh-water strata a set of marine sands and clays were deposited, and upon these thick beds of pure white earthy limestone, all belonging to the Cretaceous period. The lowest of these formations is known as the Lower greensand (*s d*, fig.10, p. 78); then comes the clay of the Gault, above which lies the Upper greensand. Then resting upon the Upper greensand comes the vast mass of Chalk (*c*, fig. 10, and No. 11, fig. 6), the upper portion of which contains numerous bands of interstratified flints which originally were mostly sponges, since silicified. The chalk strata, where thickest, are from one thousand to twelve hundred feet in thickness. The upheaval of the Chalk into land brought this epoch to

an end, because those conditions that had contributed
to its formation ceased in our area, and, as the uppermost member of the Upper Secondary rocks, it closes
the record of Mesozoic times in England.

This brings us to the last divisions of the British strata
of which I shall speak in this lecture. These were
deposited on the Chalk, and are termed Eocene formations (No. 12, fig. 6, p. 55). At the base they consist of
marine and estuary deposits, known as the Thanet sand
and Woolwich and Reading beds. These lie below the
London Clay and form the outer border of the London
basin. The same strata are found in the Isle of Wight,
and in part constitute the Hampshire basin. In the
Woolwich and Reading beds we have in places the same
kind of alternations of fresh-water and marine shells
that I mentioned to you as occurring in the Wealden
and Purbeck strata; but with this difference, that
though the shells belong mostly to the same genera,
they are altogether of different species—the old fresh-
water life is replaced by the new. Upon the London
Clay, which is a marine formation, were deposited the
Bracklesham and Bagshot beds. Upon these were
formed various newer fresh-water strata occasionally
interbedded with thin marine bands, the whole evidently accumulated at the mouth of another great river.
I may mention that the word Eocene was first used
by Sir Charles Lyell tc express the dawn or begin-

ning of recent life, or of that kind of life that exists
in the world at the present day. It is applied to
all the members of the lower Tertiary strata.

Now I think I have given you an idea of the series
of the larger and more solid geological formations that
are concerned in producing the physical structure of
England, and I will now endeavour to show you, by
the help of the diagram fig. 6, the part that these
formations play in producing the scenery of the
country.

First, then, in the far west, in Devon and Cornwall,
and in Wales, also in the north-west in Cumberland,
and in the Pennine chain which stretches from Northumberland to Derbyshire, we have what forms the
mountainous and more hilly districts of England and
Wales.

In Wales the country is essentially of a mountainous
character; and the middle part of England, such as
Staffordshire, Worcestershire, and Cheshire, may be
described as flat and undulating ground sometimes
rather hilly. But as a whole, these midland hills are
insignificant when considered upon a large scale, for
when viewed from any of the mountain regions in the
neighbourhood, the whole country below appears like
a vast plain. To illustrate this let us imagine any
one on the top of the granitic or gneissic range of the
Malvern Hills (*g*, fig. 6), which have something of a

mountainous character, and let him look to the west:
then, as far as the eye can reach, he will see hill after
hill stretching far into Wales (1 to 3, fig. 6); and if he
cast his eye to the north-east, he will there see what
seem to be interminable low undulations, almost like
perfect plains; while to the east lies a broad flat
(6 to 8) through which the Severn flows, bounded by
a flat-topped escarpment (9) rising boldly above the
plain, formed of the Oolitic formations which constitute so large a part of Gloucestershire. These, as the
Cotswold Hills, form a high table-land, overlooking on
the west a broad plain of Lias clay and of New red
marl. This Oolitic escarpment stretches in a more
or less perfect form from the extreme south-west of
England northward into Yorkshire; but it is clear
that the Oolitic strata were not originally deposited
in the scarped form they now possess, but once spread
continuously over the plain far to the west, and in all
probability only ended where the Oolitic seas washed
the land formed by the more ancient disturbed Palæozoic strata. Indeed I firmly believe that the Lias
and Oolites entirely surrounded this old land, passing
westwards through what is now the Bristol Channel
on the south, and the Dee and Mersey on the north.
They have only a slight dip to the south-east, and
great denudations having taken place, a large part
of them, miles upon miles in width, has been swept

away, probably partly by marine denudation; and thus it happens that a bold escarpment, once—in part at least—an old line of Coast cliff, overlooks those central plains of England, from which so vast an extent and thickness of Lias and Oolite have been removed.

An inexperienced person standing on the plain near Cheltenham or Wotton-under-edge would scarcely expect that when he ascended the Cotswold Hills, from 800 to 1,200 feet high, he would find himself on a second plain (9, fig. 6); that plain, however, being a table-land, in which here and there deep valleys have been scooped, chiefly by the aid of running water.* If you go still farther to the east, and pass in succession all the outcrops of the different Oolitic formations (some of the limestones of which form minor terraces), you come to a second escarpment (11, fig. 6) formed of the Chalk, which in its day also spread far to the west, covering somewhat unconformably the half-denuded Oolites, till it also abutted upon the ancient land formed of the Palæozoic strata of Wales. This also has been partly denuded, and so we have another great feature, in a bold escarpment of chalk which stretches from the south-west of England into Yorkshire.

The Eocene strata, which lie above the Chalk, in their day also extended much farther to the west, because

---

* Such valleys are necessarily omitted on so small a diagram, and the minor terraces on the plain, especially such as 7, are exaggerated.

here and there at the extreme edge of the escarpment
of chalk you find outlying fragments of them. The
proof of this original extension westward is shown in
the following diagram.

Fig. 7.

1. Chalk.   2. The main mass of the Eocene beds.   2'. Outlying
patch of Eocene beds near the edge of the escarpment.

It is impossible that these outliers could have been
originally deposited on this edge of chalk and not on
the other strata that lie west of the escarpment, and
therefore they originally extended westward, and with
the Chalk, have been denuded backwards, till they
occupy their present area. But the Eocene beds
being formed of soft strata—chiefly clays and sands—
though they make undulating ground, form no bold
scenery, but rest upon the table-land or in depressed
areas somewhat in the manner shown at 12 in fig. 6.

Such is the general manner in which part of our
country has attained its present form. The whole of
the west of England, that is to say, of Wales, and part
of the north, consists of Palæozoic strata, viz. : Cambrian,
and Silurian with all its igneous interstratifications,
Old red sandstone, the Carboniferous series, and the

Permian rocks. All these have been exceedingly disturbed and extensively denuded. They are formed of beds of variable hardness—some of them being of a slaty character, and others of masses of exceedingly hard igneous rocks, which attain in some instances a thickness of two or three thousand feet. You will, therefore, easily understand how it happens that with disturbed and contorted beds of such various kinds, those great denudations which commenced as early as the close of the Lower Silurian period, and have been continued 'intermittently ever since, through periods of time so immense that the mind refuses to grapple with them,—you will, I repeat, easily see how the outlines of the country have assumed such exceedingly varied and often rugged outlines as those which Wales, Cumberland, and in a less degree parts of Devon and Cornwall, now present.

But I have said that the Secondary and Lower Tertiary strata have not been disturbed nearly to the same extent as the Palæozoic or Primary formations anywhere in England. Though occasionally traversed by faults, yet with rare exceptions most of the strata have been elevated above the water without any bending or contortion on a large scale. What has chiefly taken place was a slight uptilting of the strata to the west, which, therefore, all through the centre of England, dip as a whole slightly but steadily to the east and south-east. This is evident from the circumstance that on the Cotswold hills the lowest Oolitic formation (Inferior Oolite, No. 9, fig. 6) forms the western edge of the Table-land, while, in spite of a few minor escarpments that rise on the surface of the upper plain, the higher Oolitic beds that dip below the Cretaceous strata are at a lower level than the Inferior Oolite at the edge of the plateau, as shown in fig. 6.

The result, then, of the great disturbance and denudation of the Palæozoic strata, and of the smaller disturbance and denudation of the Secondary rocks, is, that the general features of England present masses of Palæozoic strata forming a group of mountains in the west, then certain undulating grounds and plains composed of New red sandstone and Lias clay, and then two great escarpments, the edges of table-lands, which rise in some places to a height of more than a thousand feet; the western one being formed of Oolitic, and the eastern of Cretaceous strata, which, in its turn, is overlaid by the Eocene series, comprising the London and Hampshire basins. See fig. 6.

Again, if we construct a section from the Menai Straits, across Snowdon and over the Derbyshire hills to the east of England, the arrangement of the strata may be typified in the following manner (fig. 8 and Map, line 8). In the west, rise the older disturbed Palæozoic strata, Nos. 1 to 3, which form the mountain region of Wales. On the east of these lies an upper portion of the Palæozoic rocks, 4, consisting of Carboniferous rocks less disturbed than the underlying Silurian beds on which they lie unconformably. Then in Cheshire, to the east of the Dee, lie the great plains of New red sandstone 6, and these beds form plains because they consist of strata that have never been much disturbed, and the beds of which are soft and easily denuded. Then more easterly, from under the flat strata of New red sandstone, the disturbed Coal-measures again rise, together with the Millstone grit and Carboniferous limestone forming the Derbyshire hills, 4'. These strata dip first to the west, underneath the New red sandstone, and then roll over to the east, forming what is called an anticlinal curve, the limestone being in the centre, and the Millstone grit on both sides dipping west and east; and above the Millstone grit come the Coal-measures, also dipping west and east. Together they form the southern part of the Pennine chain which stretches northward into Northumberland, and is

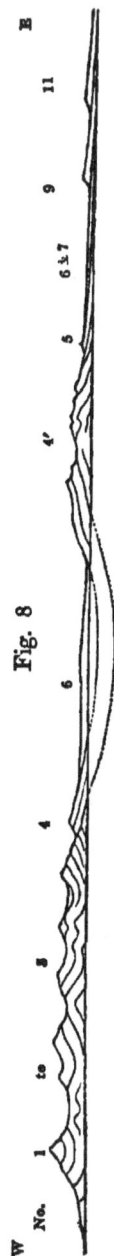

also directly united to the Cumberland mountains on the west. Then upon them, very little disturbed, dipping easterly but at low angles, we have, first, a low escarpment of Magnesian limestone 5, then the New red sandstone and Lias plains 6 and 7, which are covered to the east by the Oolite 9, forming an escarpment, the latter being overlaid by the Chalk 11 ; but in this case, the Oolitic strata being much thinner, do not form the same bold table-land that they do in Gloucestershire and in the more southern parts of England. As shown in the diagram (fig. 8), the Cretaceous rocks also rise in a tolerably marked escarpment.

Further north the same grand general features prevail. If a section were drawn across England from the Cumberland mountains to Bridlington Bay, the effect would be much as if the Lancashire and Derbyshire Carboniferous hills were added on to the east of North Wales, without the intervention of the New red sandstone and marly plains of Cheshire, for all across the Palæozoic rocks from Whitehaven into western Yorkshire it is either mountainous or hilly. Then come the New red and Liassic plains, above which rises the great Oolitic escarpment, that stretches from the Humber to the coast south and east of the Tees, and there the Oolitic series, becoming harder and thicker, attain great importance, and rise into scarped hills, as bold as the Cotswolds.

If we examine the country farther to the south-east, in the Weald of Kent and Sussex, we generally find a plain, bounded by chalk hills on the north, south, and west, while the plain itself surrounds a series of undulating hills in the centre. The whole of this Wealden area, in fact, forms a great amphitheatre, on the outside rim of which the Chalk rises in exceedingly bold escarpments, forming what are known as the North and South Downs. On the east it is bounded by the sea. There cannot be any doubt but that the Chalk and the underlying formations of Upper greensand, Gault, Lower greensand, and Weald clay originally extended fairly across all the area of the Weald for a breadth of from twenty to forty miles from north to south, and nearly eighty from east to west. This vast mass, many hundreds of feet thick, has been swept away, according to the general opinion, by the wasting power of the sea, but I believe to a great extent also by atmospheric agencies: so much so, indeed, that I am convinced that all the present details, great and small, of the form of the ground are due to the latter. The result of this is the great oval escarpment of Chalk surrounding the Wealden area, rising steeply above the plain, which is composed of strata termed the Weald clay, from beneath which the Hastings sands crop out, forming a central nucleus of hilly ground, in the manner shown in the following diagram, the height of which

is prodigiously exaggerated so as to bring the features
prominently before the eye.

Fig. 9.

N   s

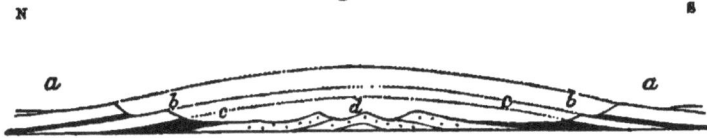

a

a   a

a a Upper Cretaceous strata, chiefly Chalk, forming the North and
South Downs; *b b* minor escarpments of Lower greensand; *c c* Weald
clay, forming plains; *d*, hills formed of Hastings sand and clay. The
Chalk, &c. once spread across the country, as shown in the dotted lines.

Let us endeavour to realise how such a result may
have been brought about. The prevalent idea that the
Wealden area once formed a vast oblong bay, of which
the Chalk hills were the coast cliffs, is exceedingly
tempting; for, standing on the edge, for instance, of
the North Downs near Folkstone, and looking south-
west across the Romney marshes, it is impossible not to
compare the great flat to a sea overlooked by all the
bays and headlands, which the winding outlines of the
Chalk escarpment are sure to suggest. And in less
degree the same kind of impression suggests itself
wherever one may chance to stand on the edge of the
chalk Downs all the way from Folkstone to Alton and
Petersfield, and from Petersfield to Eastbourne. For
years, with others, I held this view; but of late I have
begun to feel that it is not easily tenable, though it is
perhaps not very easy completely to disprove it, especially

to those who have long been accustomed to, and have never before doubted, the commonly received hypothesis.

If the Wealden area were lowered into the sea just enough to turn the Chalk escarpments into cliffs (see Map and fig. 10), we should have the following general results. Let the line $a\,b$ represent the present sea level, and the lines $s\,s\,s$ the level of the sea after depression; then so far from the area presenting a wide open sea, where heavy waves could play between the opposite Downs, we should have an encircling cliffy coast of chalk c; the base of which, unlike all coasts, is at very unequal levels. This land would be formed of two narrow strips of country, one on the south at least 60, and the other on the north not less than 100 miles long, both of which would project eastward from the Chalk of Hampshire, and form what we now call the North and South Downs. These hills generally rise high above the Eocene strata that skirt them on the north and south, which under the supposed circumstances would be covered by sea, while the scarped cliffs of Chalk, as shown on the diagram, would overlook a sea-covered plain of Gault g; outside of which, near the shore, would be a long ridgy island of Lower greensand $s\,d$, which at present, round part of the country, rises into an escarpment in places higher than the Downs themselves. Then again there would be a space of sea where the flats of Weald clay w now lie;

Effects of Submergence.

Fig. 10.

Diagram showing the general effect of a partial submergence of the Weald.   Map, line 10.

c c Chalk and Upper Greensand forming North and South Downs.
g   Gault generally forming a plain.
s d Escarpments of Lower Greensand.
w   Weald clay generally forming a plain.
h h Central Hills of Hastings sand.

inside of which would lie an island, or rather group of islands, formed of the Hastings sand series *h h*. This form of ground would certainly be peculiar, and ill adapted for the beating of a powerful surf, so as to produce *on one side only*, the cliffy escarpment that forms the inner edge of the oval of Chalk. Further, if the area had been filled with sea, we might expect to find traces of superficial marine strata of late date, as in other parts of England, scattered across the surface between the opposite downs *c c*. But none of these traces exist. On the contrary, the underlying strata of the Cretaceous and of the Wealden series everywhere crop up and form the surface of the ground, except where here and there near the Chalk escarpments they are strewn with a few flints, or where *they are covered by fresh-water sands, gravels, and loams of the ancient rivers of the country.*

I suspect, therefore, that the form of the ground in the Wealden area which has been attributed to marine action has been mainly brought about by the running water of rivers. One great effect of marine denudation when prolonged over periods of enormous length, is to produce extensive plains like the line *b b*, fig. 14, p. 140 ; for, the result produced by the wasting power of breakers is to *plane off* as it were the asperities of the land, and reduce it to its own average tidal level. Suppose the curvature of the various formations across the Wealden area to be restored by dotted lines as in the following figure, No. 11, which is very nearly on a true scale. Then let the upper part of the curve be planed across in the manner explained in a subsequent lecture, p. 141, the newly-planed surface, slightly inclined towards the interior, being represented by the line *p p*. Against this line the various masses of the Hastings sand *h h*, Weald clay *w*, the Lower greensand *s*, the Gault *g*, and the Chalk and Upper greensand *c*, would crop up. Then I believe it to be possible, and even probable, that by aid of the running water of streams, large parts of these strata might be cut away, so as to produce in an immense length of time the present configuration of the ground. If it were not so we would expect that the rivers of the Wealden area should all flow out at its eastern end, where the ground is now low, and looks out upon the sea, and towards which the long plains of Gault and Weald clay directly lead. But this, except with certain rivulets, is so far from being the case, that some streams rise close to the sea coast and flow westward. If such a plain as *p p* once existed, it is easy to understand how *the rivers might in old times have flowed from a low central watershed to the north and south across the top of Chalk at elevations at least as high as, and perhaps even a little higher than the present summit-levels.* Then, as by their action the general

Fig. 11.

Diagramatic section across the Weald, with the anticlinal curve restored as high as the Chalk. Map, line 11.

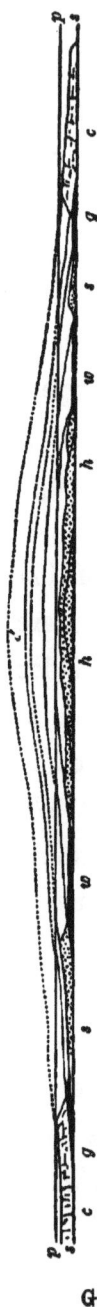

s s The level of the present sea.
p p The level of the old plain of Marine denudation after the dotted parts had been *planed* or denuded off.
c c Chalk, *g g* Gault, *s s* Lower Greensand, *w w* Weald clay, *h h* Hastings sand.
Part at least of the Eocene beds probably lay above the curved strata of Chalk marked in the dotted line *c'*.

G

level of the *inner* country was being partly reduced,
they would cut north and south channels through the
chalk downs as we now see in the Stour, the Medway,
the Dart, the Mole, the Wey, which run athwart the
North Downs, and the Arun, the Adur, the Ouse, and
the Cuckmare, which, through gaps in the South Downs,
flow south. On any other supposition it is not easy
to understand how these channels were formed, unless
they were produced by fractures or by marine denudation, of neither of which is there any direct proof.*
Through most of these gaps no known faults run of any
kind, and the whole line of the Chalk is singularly
destitute of fractures. We get a strong hint of the
probability of the truth of this hypothesis of denudation in the present form of the ground. Thus after
the formation of the marine plain *p p*, the Chalk being
comparatively hard has only been partly denuded, and
now stands out as a bold escarpment in the Downs. The
soft clay of the Gault has been more easily worn away,
and forms a hollow or plain. The Lower greensand,
full of hard bands and ironstone, more strongly resisting

---

* This kind of argument was first applied by Mr. Jukes to explain
the *behaviour* of some of the rivers of Ireland, and he then said that it
might possibly apply to the Weald.—*Geol. Journal*, 1862, vol. xviii. p. 378.
The rejection of the old hypothesis of the common marine denudation of
the Weald has been gradually forcing itself on Mr. Drew, Mr. Foster,
and other officers of the geological survey who mapped the area, and
in these remarks, in some degree, I express their opinions as well as
my own.

denudation, forms a second range of scarped hills over-looking the more easily wasted Weald clay, which forms a second broad plain, from under which rises the harder subdivisions of the Hastings sands forming the un-dulations seamed by brooks of the central hills of Ash-down Forest, and other places. The absence of flints over nearly the whole of the Wealden area, excepting near the Downs, is easily explained by this hypothesis, *for the original marine denudation had removed all the Chalk, except near the margin, long before the river denudations which commenced the escarpments began.*

Given sufficient time, I see no difficulty in this result. But the question arises, how much time in a geological sense can be given?

It is well known that, excepting for a few feet close upon the coast, this southern part of England was not depressed beneath the sea during the great icy period of the Drift. It has, therefore, been above water for a very long time. On the edge of the North Downs there are certain fragmentary outliers described by Mr. Prestwich. These by some persons have been supposed to be outliers of the lower Eocene strata called the Woolwich and Reading beds, but Mr. Prest-wich considers them to belong to the lowest part of the Crag.

If they belong to any part of the Eocene series, then the denudation of the Weald that produced its present

'form may have been going on ever since the close of
the Eocene period, that is to say, all through the
Miocene and subsequent epochs. Those who are ac-
quainted with Continental geology will realise the
meaning of this when they consider that it implies a
lapse of time far longer than it has taken to form
the labyrinthine network of valleys cut into the great
table-land of the Rhine and Moselle, or more striking
still, to form the whole range of the Jura, and all the
lowlands of Switzerland that lie between those mountains and the Alps. On the other hand, if the outliers on
the chalk escarpment west of Folkestone be parts of
the earlier Crag beds, then the bay-like denudation of
the Weald has probably entirely taken place since that
epoch; implying another lapse of time so long, that by
natural processes alone, in rough terms, half the animal
species in the world have disappeared and been as slowly
replaced by others. This may mean little to those who
still believe in the sudden extinction of whole races of
life; but to me it signifies a period analogous to the
distance of a half-resolved nebula, so vast, that if it
were possible to express it in figures the mind would
refuse to grasp its immensity.

I have gone so far into details on the point because
the ' Denudation of the Weald ' has given rise to much
theorising by several distinguished authors, and I wish
to shew the reasons why I think that the amphitheatre-

like form of the area and the escarpment of the chalk
are not directly due to marine denudation or the beating
of sea waves, but rather, that the harder outer crust of
the chalk that once cased the anticlinal curve having
been *planed off* by marine denudation, and by subsequent elevation, a table-land having been formed, the
softer rocks below that cropped up to the surface of
this plane were then attacked by running water, and
worn away so as to form the hills and valleys of the
district including the great escarpments of the North
and South Downs.

Though the Secondary and older Tertiary strata generally lie flat or dip at low angles, yet in one instance
they have been very considerably disturbed; for on a
line which runs through the Isle of Wight and the Isle

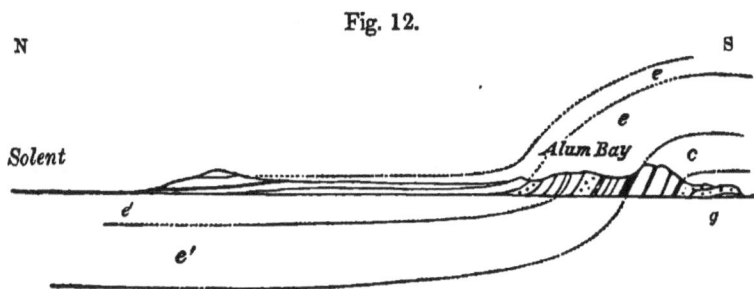

Fig. 12.

of Purbeck they stand fairly on end. Those who are
familiar with the Isle of Wight will remember that
from east to west, or from White Cliff Bay to Alum
Bay, there is a long range of Chalk hills *c*, the strata

of which dip towards the north, and are overlaid by
the older Tertiary strata *e*, that is to say the Woolwich
and Reading beds and the London clay, the Bracklesham
and Bagshot sands, and the higher fresh-water beds of
the Eocene series.

The whole pass under the Solent, as shown in the
lower dotted lines *e' e'*, fig. 12, and rise again on the
mainland in Hampshire, a considerable portion of which
is entirely composed of various sub-divisions of the
Eocene rocks.

Now these disturbed strata were originally deposited
horizontally, and after disturbance the Chalk *c* once
spread over an extensive area of Lower greensand, &c. *g*,
to the south, and the Eocene rocks *e* once spread over
the Cretaceous rocks in a curve, at a great height, as
shown in the dotted lines *e e*. Here then in our
Secondary and Tertiary rocks you get evidence of the
same kind of disturbance and denudation, of which we
have such striking proofs when we consider the structure of the country in the western and north-western·
area, which are composed of Palæozoic rocks. But in
the central part of England the Secondary and Tertiary
strata, not having been so much disturbed, have necessarily not been much denuded in height, but chiefly
backwards from west to east.

I have still, however, a few words to add respecting
the denudation of the Eocene strata. Some of these

beds in the Woolwich and Reading and in the Bagshot series consist of sands, portions of which become exceedingly hard, especially when exposed to the air. I have already said that these formations, together with the Chalk, once spread much further to the west than they do now, because outlying patches of Eocene rocks occur here and there almost at the very edge of the great Chalk escarpment, as shown in figure 7. The original continuation of both in a westward direction is shown in the dotted lines in the same diagram.

Now it so happened that when the wasting process took place that wore away both these formations from west to east, the softer clays and part of the sands of the Eocene strata were more easily removed than certain much harder portions of the sands, and the result is that over large areas, such as Marlborough Downs, great tracts of chalk are strewn with huge blocks

Fig. 13.

of tabular sandstone lying so close together, that sometimes over miles of country you may almost leap from block to block without touching the chalk on which they lie. In the above figure No. 1 represents the Chalk, and 2 the overlying Eocene clays and sands ; and the isolated blocks lying directly on the topmost beds of the chalk represent the thickly scattered masses of stone left on the ground *after the removal by denudation of other and softer parts of the Eocene strata* No. 2. Frequently they are found scattered even on the terraces of the Lower Chalk, a remarkable example of which occurs at the old British town of Avebury, near which the lower terrace of Chalk (as in the diagram) is strewn with 'grey wethers,' as they are termed, and immense masses set on end by a vanished people stand in the ancient enclosure. Sometimes even on the plains of Gault or Kimeridge clay well out to the north or west of the escarpment, as for instance at Swindon, blocks angular or half-rounded lie in the meadows, marking the immense waste to which the whole territory has been subjected long after the close of Eocene times. Besides the name of 'grey wethers,' they are known by the name of Sarsen stones, and Druid stones, and all the standing masses of Avebury and Stonehenge, popularly and probably erroneously supposed to be Druidical temples, have been left by denudation not far from the spots where they have since been erected into such grand old monuments by an ancient race.*

I think I have now stated enough to enable you to

---

* The smaller stones at Stonehenge have been brought from a distance. They are mostly of igneous origin, and are believed by Mr. Fergusson to have been votive offerings.

form a general idea of the geological phenomena which produce the leading features of the scenery of England. I might add many details respecting other portions of England, such as the relation of the Secondary rocks to the older rocks of Devon, the structure of the Malvern range and of the Mendip Hills, or of the beautiful Vale of Clwyd, in North Wales, consisting of a bay of soft New red sandstone, bounded by Silurian mountains and old limestone cliffs, and of the still larger Vale of Eden, in the North, where the mountains of Cumberland look down on an undulating ground of Permian and New red strata. But it would not add much to the general · knowledge which I wish to impress on you, viz. that England is mountainous in the west, or in Devon, Wales, and Cumberland, because of disturbance and great denudations; and that it consists of plains and table-lands in the central and eastern parts because the strata there are flat and softer, and because they have been denuded in such a manner that their western portions have been chiefly cut away and thus their edges form long escarpments.

In the next lecture I shall have something to say about the softer coverings of this hard skeleton.

LECTURE IV.

THE MIOCENE AND PLIOCENE TERTIARY STRATA. GLACIAL
PHENOMENA ; AND ORIGIN OF CERTAIN LAKES.

WE now come to the Middle and Upper Tertiary strata,
the first of which consists of the Miocene beds. Their
position is shown on the geological scale at page 20,
above the upper Eocene formations. They, however,
play such a very unimportant part in the *physical
geology* of the mass of our country, that I shall dismiss
them in a very few words.

The Miocene beds are only known in Britain in the
island of Mull, one of the Western islands of Scotland,
where there are certain strata of shale, interstratified
with beds of basalt and volcanic ash, first described
by the Duke of Argyll, and known to be of Miocene
date, because of the plants which occur in them being
all distinct from any living species, and many the
same as those well-known to be of Miocene age, in

Bohemia, on the banks of the Rhine, in Switzerland, and in other places where Miocene formations are well developed.

In the south-west of England, in the neighbourhood of Dartmoor, at a place called Bovey Tracey, in a flat area ten miles long by two miles wide, there are also found beds of Miocene clay interstratified with bands of imperfect lignite; and of late these beds, the age of which was for long a puzzle, have been investigated through the liberality of Miss Burdett Coutts, who paid all the expenses to enable a gentleman in that neighbourhood to examine the nature of these strata because there were no commercial works there of sufficient importance to develop them; and it required digging in order to enable any one to arrive at just conclusions as to the nature of the strata. The result was, from an examination of the fossil plants by Professor Heer of Zurich, that they were also found to be of Miocene date; and it is an important fact, that similar plants are not only found here and there in Scotland, Ireland, and England, in Bohemia, and on the Rhine, but many of them also occur in Iceland, and in North America and Greenland beyond the Arctic circle. The meaning of this is not yet understood; for many of the plants are of a nature that seem to bespeak a warmer climate than that of the British Islands at the present day, and it is besides difficult to see how such plants could grow in Arctic regions, where the stimulus of light is wanting during half the year. This is one of those things which we cannot explain, and about which we are waiting for light.

Above the Miocene beds come the Pliocene strata; that is to say beds still newer in the series, and these in the lower part consist of subdivisions generally known as the Crag (a workman's term). At the base lies the Coralline Crag, and above this lies the Red Crag, which in some places is not conformable to the Coralline Crag below, showing that an interval marked by denudation elapsed between the deposition of the strata. Much newer than the Red Crag is the Norwich or Mammaliferous Crag, differing in many details from either of the others. All of these occupy certain minor patches in Norfolk and Suffolk; and here and there on the borders of the Weald, in some situations on the very top of the Chalk Downs, there are small patches of sand which are provisionally placed by some geologists on the horizon of part of the Crag. But this is doubtful. The lower, or Coralline Crag, contains 51 per cent. of species of shells still living in the present seas. The Red Crag has a still larger proportion of existing species, showing that it approaches nearer to our own day, not only because it lies above the Coralline Crag, but also because it contains a greater per-centage of living species mixed with the fossil remains; and along with the shells which form the chief mass of that formation, there are found the bones of a few species of Mammalia, some belonging to the sea, and others to the land. Those of marine origin are cetacean, as the whale; and along with these also occur the remains of the Mastodon, a remarkable kind of Elephant, with teeth which differ very much from that of the true elephant in certain particulars, but these details it is not my object now to explain. Then, in the Mammaliferous crag of newer date, we have a larger proportion of Mammalian remains, hence its name. In it have been found the bones of the Mastodon, also of an extinct Elephant and other large mammalia, and of the ass, the beaver, and a great number of other smaller animals. I mention these things, not because these formations play an important part in British physical geography, as they are generally so far buried under superficial strata of one kind or another that they require to be looked for, and thus do not at all affect the scenery, but to show you the kind of changes in physical geography that our country, in comparatively late times, must have undergone. If it has undergone these changes in late times, far greater are the numbers and the kinds of changes that it must have undergone in periods that went before, of which the records are often entirely lost.

We are not of necessity to consider Great Britain as

having always been an island during and between the periods that I have already described. It is an accident that it is now an island; and it has been an island or islands probably many times before, and in many shapes; and when you consider that we have here two epochs of the Crag, both containing remains of great terrestrial animals, you will see that it must have been joined at times to the main-land, for otherwise these great terrestrial animals could not have found their way into our area. When I describe other periods, still later than the Crag, we shall be able to understand a little more definitely the precise kind of changes that our land in latter days has undergone.

Younger than the Crag there are certain other minor deposits, portions of which are scattered here and there throughout England. One of the most remarkable of these is the 'Forest Bed,' lying underneath the glacial drift on the shore at Cromer in Norfolk. This bed has been traced for miles between high and low water mark, and contains numerous erect stumps of Scotch fir, spruce, yew, alder, oak, &c., together with remains of three elephants, *E. meridionales*, *E. antiquus*, and *E. primigenius* (the Mammoth), a rhinoceros, hippopotamus, horse, pig, and other mammalia; and the whole speaks of a past physical geography, at least during part of which, with a mild climate, our country may have been joined to the Continent.

Later still there are other small formations, important in themselves and definite to those who study fossil remains, but like the Cromer beds, as they scarcely affect the features of the country, I shall say nothing about the causes that brought about a patch here and a patch there of gravel or loam, in which we find relics of the hippopotamus, the rhinoceros, and other mammalia.

But I must now describe a remarkable episode in the latest Tertiary (or, as some authors call them, the Post-Tertiary) times, known as the *Glacial epoch*, said to be altogether of later date than the Cromer 'Forest Bed,' and certainly of earlier date than some of the patches just alluded to. This formation has left its traces universally over the whole northern half of the northern, and also over a large portion of the southern hemisphere; and I hope I shall be able to describe the history of that period, as it affects the scenery of Britain, with something like tolerably accurate detail. Before doing so, however, I must lead you into Switzerland, and show you what kind of effect is being produced there by the ice of the present day, and afterwards into Greenland, and show you what takes place there, and then by the knowledge thus gained, I shall be able to bring you back into our own country, and explain what took place here in that icy episode which is so far distant in time, but which, by comparison with the more ancient periods, almost approaches our own day.

Now the first thing I have to do, is to describe what a glacier is. In any large and good map of Switzerland you will see certain white patches here and there on the higher mountain ranges of the Alps. The highest mountain in the Alps, Mont Blanc, rises more than fifteen thousand feet above the sea, and there are other mountains in this great chain which approach that height, ranging from 12,000 to 15,000 feet. The mean limit of perpetual snow upon the Alps is 8,500 feet above the level of the sea. Above that line, speaking generally, the country is mostly covered with snow, and in the higher regions it gathers on the mountain slopes and in the larger recesses, and by force of gravity it presses downwards into the main valleys; where, chiefly in consequence of the immense pressure exerted by the weight and movement of this accumulated mass, the snow year after year is converted into ice. Without entering on details, it is enough if I now state that this is proved by well-considered observations made by the best observers of the icy phenomena of the Alps. Still accumulating, year upon year, by degrees this ice slides down the valleys, and is often protruded in a great tongue far below the limits of perpetual snow ; for some glaciers descend as low as from *three* to *four* thousand feet or thereabouts above the level of the sea, whereas the limit of perpetual snow is 8,500 feet. Now I will not enter into all the details of the structure of glaciers, because that will not help us in the special investigation we have now in view; but I will describe to you what are the effects produced by a glacier in the country over which it slides, and various other glacier-phenomena affecting the scenery of the Alps, and therefore affecting the scenery of our own country in past times when glaciers existed here, and still affecting it in the relics they have left.

A glacier slides more or less rapidly, according to the mass of ice that fills the valley, and also according to the greater or less inclination of the slope, for in this respect it behaves very like a river. If you have a vast body of water like the Mississippi flowing down a broad valley, although the slope of the valley may be very gentle, still the river flows with great rapidity in consequence of the greatness of the body of water; so if you have a mass of ice, which represents the snow-drainage of a large tract of country, *covered with perpetual snow,* then the *glacier flows with a rapidity proportionate to the mass of ice,* and that rate of progress is modified, increased, or diminished, in accordance with the fall and width of the valley, so that when it is steep, the glacier flows comparatively fast, and when the angle at which the valley slopes is small, it flows with comparative slowness.

All glaciers are traversed by cracks which are termed

H

crevasses. Now the mountain peaks that rise above the surface of a glacier in some cases are so steep that the snow refuses to lie upon them, even when they may happen to be above the limits of the average line of perpetual snow, so that masses of rock are always being severed by atmospheric disintegration, and falling from the slopes they find a temporary resting place on the surface of the ice at the margin of the glacier, and, as it were, float upon its surface in long and continuous lines; for the motion of a glacier is so slow, that the quantity of stones that fall upon its surface is sufficiently numerous to keep up a continuous line of blocks, earth and gravel, often of great width. These stones, when two glaciers combine to form one great stream of ice, as in the glacier of the Aar, at a ·certain point meet and form one grand line running down the centre of the glacier. These are termed *moraines,* and at length all of this material that has not fallen into crevasses floats on as it were to the end of the glacier, and is shot into the valley at the end of the ice-stream, frequently forming large mounds, known as *terminal moraines.* .

Beneath every glacier water is constantly flowing, caused by the melting of the ice both below and on the surface of the glacier, and also in some cases to a less degree, by springs that rise in the rocks below the ice. In the various parts of glaciers, where crevasses are not numerous, you frequently find large brooks so wide that you cannot leap across them, and you may have to walk half a mile before you find a passage; but in all the glaciers that I have seen, long before you reach their lower end, all the surface water has found its way to the bottom of the ice. The water therefore that runs from the end of a glacier very often emerges from an ice-cavern as a large ready-made muddy river, which carries away the moraine rubbish that the glacier deposits at its lower end, in some cases almost as fast as it is formed; perhaps I might rather say as slowly as it is formed, because if you go day after day you might see scarcely any difference in the detail of certain moraines, though in time, when favourably placed to be worked upon by water, stones of moderate size that have been shed from the ice are carried by the river down the valley. In other cases, however, it happens that from various circumstances moraines are preserved from destruction, and form permanent features in the scenery.

Now I have something special to say about moraine-stones before I describe the glacial phenomena of our own island. When an immense weight of ice, in some cases thousands of feet in thickness, forming a glacier, passes over solid rocks, by the pressure of the moving mass, the rocks in the valley over which the glacier passes become smoothed and polished—not flatly, but in wavy lines, presenting a largely mammillated surface. Furthermore, the stones of the surface moraines frequently fall into crevasses, and the small débris and finely powdered rocks that more or less cover the surface of a glacier are also borne into these crevasses by the water that flows upon the surface; the consequence is that the bottom of a glacier is not simply bare ice, but between the ice and the rock over which it flows, there are blocks of stone imprisoned, and siliceous and feldspathic débris (chiefly worn from the floor itself), which may be likened to emery powder. The result is, that let the rock be ever so hard, it is, in time, polished almost as smooth as a sheet of glass, and this polished surface is scratched and grooved by the coarser débris that, being imprisoned between the ice and the rocky floor, is pressed along in the direction of the flow of the ice. By degrees, grooves and deep furrows are thus cut in the rock over which the ice passes.

But the stones that are imprisoned between the ice and the rocky floor not only groove that floor itself, but in turn become scratched by the harder asperities of the rocks over which they are forced; and thus it happens that many of the stones of moraines are covered with straight *scratches*, often crossing each other irregularly, so that we are able by this means to tell, independently of the forms of the heaps, whether such and such a mass is a moraine or not.

These indications of the rounding, smoothing, scratching and grooving of the rocks in lines coincident with the flow of the glacier, together with old moraine heaps and scratched stones, are so characteristic of all glaciers, that by this means we are able to detect the important fact that the Swiss glaciers were once of far larger dimensions than they are now, and that they have gradually retreated to their present limits. Far below the present ends of the Swiss glaciers, 50 or 80 miles farther down the valleys, we find all the signs I have described, and others besides, frequently as marked as if the glacier had only left the rocks before the existing vegetation began to grow upon their surfaces.

Such being the case in Switzerland, where we have been able to study the action of glaciers in detail, we have next to enquire, is there anything further to learn in regions where glaciers are on a far greater scale? Those who have read the descriptions of navigators will be aware that in Greenland, the average ice-line, as a whole, descends lower and lower as you go northward, till in the extreme north the whole country is one universal glacier. The same universal covering of ice is found in that southern land, discovered by Ross, and known as Victoria Land, where the mountains rise some of them ten, twelve, and fourteen thousand feet above the sea, and except here and there, where the cliffs are very steep, the country is covered with a coating of thick ice. In ,Greenland, where the coast happens to be high and steep, the glaciers break off at the top of the cliffs and fall in shivered icebergs into the sea, but when valleys fairly open into the sea then it frequently happens that prodigious glaciers push their way across the land out to sea; and are in certain cases 12 or 14 miles across at their ends. In the extreme north the glacier has been described as proceeding out to sea, and forming a continuous cliff of ice as far as the eye can reach, far outside the true rocky coast. Some of these vast glaciers have been estimated as being at the very least 3,000 feet in thickness ; and great masses of ice breaking away from their ends, form icebergs, which frequently laden with moraine rubbish, just like that which covers the glaciers of Switzerland, float out into the Greenland seas, and are carried south by a current, along the coast of North America. Some of these bergs are known to float south beyond the parallel of New York, and they have even been seen off the Azores. Melting by degrees as they come into warmer climates the stony freight is scattered abroad, here and there over the bottom of the Atlantic, which thus becomes strewn with erratic blocks, and other débris borne from far northern regions.

I shall now apply these remarks in our own island, and having ascertained what are the signs by which a glacier may be known, I shall show that a large part of the British islands has been subjected to *glaciation,* or the action of ice.

Those who know the mountains of the Highlands of Scotland remember that though the weather has had a powerful influence upon them, rendering them in places rugged, jagged, and cliffy, yet, notwithstanding this, their general outlines are often remarkably rounded, flowing in greater and smaller mammillated curves; and when you examine the valleys in detail you also find that in their bottoms and on the sides of the hills, the same mammillated structure frequently pre-vails. These rounded forms are known to those who specially devote themselves to the study of glaciers by the name of *roches moutonnées,* a name now in general use in England, because it happened that in Switzer-land glaciers were first described by authors who wrote in French. Ice rounded rocks are exceedingly com-mon in many British valleys, and not only so but the very same kind of grooving and striation, so emi-nently characteristic of the rocks in the Swiss valleys, also marks those in the Highlands of Scotland, Cum-berland, Wales, and other districts in the British islands. Considering all these things, geologists, led twenty-three years ago by Agassiz, have by degrees almost universally come to the conclusion that a very large part of our island was, during ' the glacial period,' covered, or nearly covered, with a coating of thick ice, in the same way that the north of Greenland is at present; so that by the long-continued grinding power of a great glacier or set of glaciers nearly universal over the northern half of our country and the high ground of Wales, the whole surface became *moulded by ice*; and the relics of this action still remain strongly impressed on the country, to attest its former power.

It might be unsafe to form this conclusion merely by an examination of such a small tract of country as the British islands, but when we consider the great Scandinavian chain, and the north of Europe generally, we find that similar phenomena are common over the whole of that area; and in the North American continent, as far south as latitude 38° or 40°, you find, when you remove the soil or the superficial covering of what is called drift, and get at the solid rock beneath, that almost everywhere it is smoothed and polished, and covered with grooves and striations similar to those of which we have experience among the glaciers of the Alps. I do not speak merely by common report in this matter, for I know it from personal observation, both in the Old World and the New. We know of no power on earth, of a natural kind, which produces these indications except ice, and therefore geologists are justified in attributing them, even on this great continental scale, to its action.

You will presently see that this conclusion is forti-
fied by several other circumstances. Thus, in the Alps,
there is evidence that the present glaciers were once on
an immensely larger scale than at present. The proof
as usual lies in the polished and grooved rocks far
removed from the actual glaciers of the present day,
and in numerous moraines on a scale so immense that
the largest forming in the Alps in our time are of mere
pigmy size when compared with them. The same kind
of phenomena occur in the Himalayah, the Andes,
and in almost every northern mountain chain or cluster,
great or small, that has been examined critically, and
therefore there can be no doubt that at a late period of
the world's history an extremely cold climate prevailed
over much larger tracts of the earth's surface than at
present, produced by some cause about which there are
many vague guesses, but which no one has yet explained.

It was at this period that a great part of what is now
the British islands was covered with ice. I do not say
that they were islands at that time, and I think they were
not islands, but probably united with the Continent, and
the average level of the land may then have been much
higher than at present, chiefly by elevation of the
whole and partly because it had not suffered so much
degradation ; but whether this was so or not, the mountains and much of the lowlands were covered with a

universal coating of ice, probably as thick as that in the
north of Greenland in the present day. While this
large ice-action was going on, a slow submersion of the
land took place ; and as it sank the glaciers, descending to the level of the sea, deposited their moraine
rubbish there. Gradually the land sunk more and
more, the cold still continuing, till this country, previously united to the continent of Europe, became a
group of icy islands, still covered with snow and small
glaciers, which descended to the sea, and broke off in
icebergs. These floating south deposited their stony
freights as they melted. The proof of this is to be
found in the detritus which covers so much of Scotland and two-thirds of England, composed of clay and
gravels mixed with stones and great boulders, many
of which are scratched, grooved and striated, in the
manner of which we have every-day experience in the
glaciers of Switzerland, Norway, and Greenland.

Much of this clay is known as the 'Till' in Scotland ;
and it was only by very slow degrees that geologists
became reconciled to the idea that this Till is nothing
but moraine rubbish on a vast scale, formed by those
old glaciers that once covered the northern part of
our country. In fact, Agassiz who held these views,
and Buckland who followed him, were something like
twenty years before their time ; and men sought to
explain the phenomena of this universal glaciation by

every method but the true one. Mr. Robert Chambers was, I think, the first after Agassiz, who asserted that Scotland had been nearly covered by glacier ice, and now the subject is being worked out in all its details; thus coming back to the old generalised hypothesis of Agassiz, which is now accepted or on the very verge of acceptance by most of the best geologists of Europe and America.

Besides the proofs drawn from the scattered boulders, we know that the country was descending beneath the sea during this glacial epoch for another reason,—that here and there in the heart of the moraine-matter of the Till, there are patches of sand and clay interbedded. The mass indeed is not stratified, because glaciers do not stratify their moraines, but the waves playing upon them, as they were deposited in the sea, here and there arranged portions in a stratified manner; and there occur at rare intervals, in these patches in Scotland,* the remains of sea-shells of species such as now occur in the far north. Here, therefore, we have another proof of that arctic climate which, in old times, came so far south.

In Wales we find similar evidence, long since described by myself,† of the sea having risen at least 2,300 feet upon the sides of the mountains, for Wales, like Scot-

* Lately described to me by Mr. Geikie, in the Scotch Till.
† See 'The Old Glaciers of North Wales.'—*Ramsay.*

land, also became a cluster of islands, round which the
drift was deposited, and great blocks of stone were
scattered abroad, floated out on icebergs, that broke
from an old system of glaciers, and melted in the
neighbouring seas. In this stratified material sea-shells
were long ago found in Caernarvonshire by Mr. Trimmer
and myself, from 1,000 to 1,400 feet above the sea.

Erratic blocks of granite, gneiss, feldspathic traps,
and of other rocks, some of which came from the
Highlands of Scotland, some from the Cumberland
mountains, some from the Welsh mountains, and some
from the farther region of the great Scandinavian chain,
were in the same manner spread over the central
counties, and the west and east of England, just like
those boulder-beds that are now being formed at the
bottom of the Atlantic from the icebergs that float south
from the shores of Greenland.

All of these marine boulder-drifts are rudely stratified when viewed on a large scale, and the clays are
often interbedded with sand and rounded gravel, but
it is remarkable that in most of the great beds of clay
that form the larger part of the formation, the stones
and boulders that stud the mass are scattered confusedly,
and frequently stand on end, like the stones in the
Moraine-Till of Scotland, and this is the case even
when associated with sea-shells, which shells prove
the true marine nature of what often looks like a mass

of heterogeneous rubbish. A great number of these shells occur in a broken state, and from Scotland to Norfolk on the coast cliffs, they may be found plentifully enough when carefully looked for. Between Berwick and the Humber, I have seen them in scores of places; the most plentiful species, as determined for me by Mr. Etheridge, being *Cardium edule, Cyprina Islandica, Venus, Dentalium entalis, Tellina, Leda oblonga, Astarte borealis,* and *Saxicana rugosa.* On the west coast by the Mersey and near Blackpool, they are equally plentiful, and far inland near Congleton and Macclesfield, I observed the same kind of broken shells 600 feet above the sea, and I understand that Mr. Prestwich observed them in the same region at a height of fully 1,200 feet. It is remarkable also what a prodigious number of ice-scratched stones occur in this drift, under such conditions, that the idea is suggested that they were marked not by glaciers, but by the agency of coast ice. As you travel south, you find that the numbers of the kinds of stones increase according to the number of formations you have passed. North of the Magnesia limestone district, the fragments to a great extent consist of Silurian Old red and Carboniferous fragments, then these become mixed with pieces of Magnesian limestone, by and by Oolitic fragments are added, and in Holderness, in places half the number of stones are of chalk generally well scratched, and often mingled with broken shells. But it is evident that the low chalk hills of Yorkshire are not of a kind to have given birth to glaciers, and therefore the scratching and distribution of the chalk-stones may have been produced by coast ice; and the same may be said of other low parts of England, both on the coast and inland where the drift prevails, the currents which scattered the ice-borne material having on a great scale flowed approximately from north to south.

But England, south of the estuaries of the Severn and the Thames, for the most part seems all this time to have remained above the waters, for not only is the country in general destitute of drift, but it is only close on the sea near Selsea and Brighton that erratic boulders of granite &c. have been found, apparently floated from the Channel Islands or from France.

After a long period of submergence the country gradually rose again, and the evidence of this I will prove chiefly from what I know of North Wales, although I could easily do the same by taking you to the Highlands of Scotland.

I shall take the Pass of Llanberis as an example, for there we have all the ordinary proofs of the valley having been filled with glacier-ice. First, then, after the great glacial epoch, the country to a great extent sunk below the water, and the drift was deposited, and more or less filled many of the valleys of Wales. When

the land had risen again to a considerable height, the
glaciers increased in size, although they never reached
the immense magnitude which they attained at the
earlier portion of the icy epoch. Still, they became
so large, that such a valley as the Pass of Llanberis
was a second time occupied by ice, and the result was,
that the glacier ploughed out the drift and loose rubbish, that more or less covered the valley. Other
cases of the same kind could easily be given, while,
on the other hand, in many valleys you find the drift
still remaining. By degrees, however, as we approach
nearer our own days, for unknown reasons, the climate
slowly ameliorated, and the glaciers began to decline,
till becoming less and less, they crept up and up ; and
here and there, as they died away, they left their terminal and lateral moraines, still in some cases as well
defined as moraines in lands where glaciers now exist.
Frequently too, masses of stone, that floated on the
surface of the ice, were left perched upon the rounded
roches moutonnées, in a manner somewhat puzzling
to those who are not geologists ; for they lie in
such positions that they clearly cannot have rolled into
them from the mountain above, because their resting
places are separated from it by a hollow; and besides
many of them stand in positions so precarious, that if
they had rolled down from the mountains they must,
on reaching the points where they lie, have taken a

final bound, and fallen into the valley below. But when experienced in the geology of glaciers, the eye detects the true cause of these phenomena, and you have no hesitation in coming to the conclusion, that as the glacier declined in size, the errant stones were let down upon the surface of the rocks so quietly and so softly, that there they will lie, until an earthquake shakes them down, or until the wasting of the rock on which they rest precipitates them to a lower level. Finally, the climate still ameliorating, the glaciers shrunk farther and farther into the heart of the mountains, until, at length, here and there in their very uppermost recesses, you find the remains of tiny moraines, marking the last relics of the ice before it disappeared from our country.

All these things give distinctive characters to our mountains, very different, I believe, from those of mountain ranges where glaciers never were. I confess, however, that *I have never seen any of the latter*, my travels having been confined to Europe north and immediately south of the Alps, and to parts of North America.

There are certain other points that materially affected the geography of our country, and that is what I believe to be the glacial origin of many of our lakes.

When glaciers descended into valleys, and deposited their terminal moraines, it has sometimes happened that when a glacier declined in size its moraine still remained tolerably perfect, with this result—that the drainage formerly represented by ice is now represented by running water, which is dammed in between the surrounding slopes of the solid mountain and the mound formed by the terminal moraine, thus making a lake. There are such lakes on the Italian side of the Alps, and there are several among the mountains of Wales. Whether there are any in Cumberland I do not know, but as yet I have seen none in Scotland dammed by the terminal moraines of common valley glaciers, although I have no doubt that they may exist in parts that I have not visited. Furthermore, sometimes on the *outer* side of these moraines we find stratified boulder-drift, showing that the old glacier descended to the level of the sea and deposited its moraine there, and breaking up, floated about as icebergs bearing boulders. By-and-by the glacier that was produced by the drainage of snow disappeared, and is now represented by water, forming a lake dammed by a moraine, outside of which lie long smooth slopes of stratified drift.

Such lakes are always on a small scale, but there are others on a larger scale, having a far more important bearing upon the physical geography of our country, and of many other countries in the northern hemisphere, and I have no doubt also in the south. The theory which I am going to propound to you is my own, and is not old. It gave rise to a considerable amount of opposition, and also to some approval, and I believe that in time it will be sure to make its way.

There is no point in physical geography more difficult to account for than the origin of most lakes. When thought about at all, it is easy to see that lakes are the result of the formation of hollows, a great proportion of which are true *rock-basins,* that is to say, hollows entirely surrounded by solid rocks, the waters not being retained by mere loose detritus. But the great difficulty is, how were these *rock-basins* made? In the first place, consider what is the effect of marine denudation. On the sea-shore, where the waves are always breaking, the effect of this and of the weathering of cliffs that rise above the waves is to waste back the land. But the sea in this case *cannot make a hollow below its own average level.* What it does, if there are any hollows there, is to fill them up with detritus, for it cannot cut them out. The consequence is, that the chief power of the sea working against the land and wasting it back, is to act as a great planing machine, wearing off the larger irregularities that rise above its level in the manner shown in the description of the first denudation of the Weald at page 80, and of South Wales at page 140.

Again, what is the effect in any country of running water? *Rivers cannot make hollows that are surrounded by rocks on all sides.* All that running water can do upon the surface is to scoop out trenches or channels of greater or less width, forming gorges or wider valleys, according to the nature of the rivers and the rocks, and the time employed in the work. If you have an inclined plane, with a long slope, gentle or steep, water will run upon it because of the slope, and, aided by atmospheric disintegration, it will cut out a channel, *but it cannot make a basin.*

Again: it may be contended that the hollows were formed by the disturbance of the rocks, so as to throw them into a basin-shaped form. But when we take such lakes as that of Geneva, the Lake of Thun, the lakes of Lucerne, Zurich, Constance, and the great lakes on the Italian side of the Alps, and examine the strata critically, we find that they do not lie in the form of basin-shaped synclinal hollows. Such depressions are the rarest things in nature; that is to say, hollows formed of a number of strata bent upwards at the edges all round into the form of a *great dish*, the very uppermost bed or beds of which shall be continuous and unbroken underneath the water of the lake. If such basins exist at all, I never saw one, though specially looking out for them in many regions, and I believe that they have been only assumed by those who have not realised the meaning of denudations on a large scale, and therefore aré apt to consider hills and valleys as the result, mainly, of disturbance and dislocation. From repeated examinations I feel assured also that the Swiss valleys generally, and the lake-valleys in particular, do not lie in gaping rents or fissures, and, indeed, after half a life spent in mapping rocks, I believe that there is no *necessary* connection between fractures and the formation of valleys.

It might, however, be said that these lakes lie in areas of special depression, made by the sinking of the land underneath each lake. Lakes are, however, so numerous in the Alps, and in the Highlands of Scotland, where they occur by the hundred, and in North America by the thousand, that I feel sure the theory of a particular depression for each lake will not hold. In the northern part of North America it is as if the whole country were sown broadcast with lakes, large and small, and great part of the country not being mountainous, but consisting of undulating flats, it becomes an absurdity to suppose that, so close together, a special area of depression was provided for each lake. The physical geology of America and of Scotland entirely goes against such a supposition, and I believe, though the subject is a little less evident, that it is equally untenable for the Alps, and the lowlands between the Alps and the Jura.* Now, when you have come to

* For details see Quarterly Journal of the Geological Society, 1862, p. 185. There are some lakes and inland seas said to occupy areas of special depression, for I by no means wish you to understand me as asserting that this theory accounts for the probable origin of all lakes.

these conclusions, you will see that it is difficult to account for the existence of hollows, composed of hard rocks, which completely inclose lakes.

If, then, we have disposed of these erroneous hypotheses for the formation of such hollows, what is left? If the sea cannot do it, nor weather, nor running water, and if the hollows were not formed by synclinal curves of the strata, and if they do not necessarily lie in gaping fissures, nor yet in areas of special depression, *the only remaining agent* that I know *is the denuding power of ice.* In the region of the Alps it is a remarkable circumstance that all the larger lakes lie in the courses of the great old glaciers—each lake in a true rock-basin. This is important, for though it is clear that the drainage of the mountains must have found its way into these hollows, either in the form of water or of glacier-ice, yet if ice had nothing to do with their formation, we might expect an equal number of lakes great and small in other regions where the rocks are equally disturbed or of like nature, but where there are no traces of glaciers. I have never observed that this is the case, but rather the reverse.

I will take the Lake of Geneva as a special example before applying the theory to our own country. This lake is about forty miles long, and in its broadest part about

Many are dammed up by drift, and in other regions some may owe their origin to causes of which I know nothing.

twelve miles wide. It lies at the mouth of the upper valley of the Rhone and directly in the course of the great old glacier, which was something like a hundred miles in length from its source to where at its end it abutted upon the Jura mountains, by 125 miles from south-west to north-east at its lower end. It is known to have been so large by the effects produced on the country over which it flowed, and also by the fact that huge blocks of gneiss, granite, and others from the Alps lie scattered on the flanks of the Jura, associated with moraine matter in such a way as to leave the former existence of that great old glacier beyond a doubt.

The Lake of Geneva where deepest, near its eastern end, is 984 feet in depth, and it gradually shallows to its outflow. By examining the sides of the mountains on either side of the valley of the Rhone, through which the glacier flowed, we are able to ascertain what was the thickness of the ice in that valley when the glacier attained its greatest size, viz. nearly 2,800 feet above the surface of the lake, or nearly 3,800 if we add the depth of the water.* By similar observations on the *Jura*, it is clear, that where the ice abutted on that range, it still maintained a thickness of something like 2,200 feet where thickest, swelled as it was by the vast tributary masses of the glacier of the Arve,

* But probably more.

and by others of smaller size that flowed down the valleys that open on the south of the Lake of Geneva. Now, consider the effect of this gigantic glacier flowing over the Miocene rocks, which in this part of Switzerland are comparatively soft, and ‚yet of unequal hardness! That mass, working on slowly and steadily, for a period of untold duration, must have exerted a prodigious grinding effect on the rocks below. Where the glacier-ice was thickest, there the grinding power was greatest, and the underlying rock was to a very considerable extent destroyed and worn away; but where at its western end near Geneva the ice was thinner by reason of the melting of the glacier, there the pressure and grinding power were less, and the waste of the underlying rock was proportionately diminished. The result was, that a great hollow was scooped out, at least 984 feet deep in the deepest part, without allowing for any sediment that may now cover the bottom, this hollow shelving up towards the present margin of the lake. When you think of this, you may at first deem it impossible, but when you compare the depth with the length of the lake and the height and weight of the ice above, and reduce all to a true scale, you will see that in a drawing the depth of the rock-basin is comparatively quite insignificant.*

* These points were illustrated by diagrams, as were all the other questions raised in this lecture.

Therefore I have been forced to the conclusion, from a critical examination of many of the lakes in and around the Alps, that their basins were scooped out by the great glaciers of the icy period; some of which were as thick or thicker than that which descended the valley of the Rhone.

Now, if we examine the maps of the northern hemisphere generally, beginning at the equator, and coming north, it is remarkable that we find very few lakes in its southern regions. As we proceed northwards into America, in latitudes 38° and 40°, the lakes begin to increase, and soon become tolerably numerous; and north of New York, towards the St. Lawrence, they become so numerous that they appear to be scattered over the whole country in every direction, and beyond this to the north and west of Lake Superior and the St. Lawrence, the whole country is dotted as thickly as lakes can be put upon the maps, and a vast number of the smaller ones are omitted for want of room. The whole of that country has been *completely* covered by ice—as the researches of geologists show.

Coming to this side of the Atlantic, and examining the Scandinavian chain on the east, where the slopes are less than on the western flank, all round the Gulf of Finland, and the Baltic, the whole country is covered with lakes, many of which, I am informed, lie in true rock-basins, while in Finland, according to Professor

Nordenskiold, they are in a glaciated country chiefly dammed in by heaps of detrital matter called Ösars.* Go into North Wales, where glaciers were common in every large valley: there we have the lakes of Llanberis, of Cwellyn, Ogwen, Llyn-y-Ddinas, Llyn-Gwynant, Llyn-llydaw, and all the minor tarns in the upper Corries. There are many celebrated lakes in parts of Ireland and Cumberland—both eminently glaciated countries— and others unknown to fame besides. Go from the comparatively southern parts of our island, and examine Scotland; in Sutherland, and the Lewes, and in other Western islands; in Inverness-shire, Perthshire, Dumbartonshire, and the Mull of Cantyre, the whole country is sown with lakes—a vast number of which I can testify lie in true rock-basins, though some may merely lie in hollows made by unequal accumulation of glacier drifts, or among the bending gravelly mounds of the 'Kaims.' And all that country, like Greenland now, was in the icy period ground by a heavy weight of slowly moving and long enduring glacier-ice, which ice I firmly believe was the scooping power that originated most of the lake scenery of our country. I go further,

---

* The Eskirs of Ireland and the Kaims of Scotland. These are common in the great valley of the Frith of Clyde, especially near Lanark and Carstairs, where they form vast elongated irregular mounds of gravel above the true glacial detritus. They inclose lakes and peat-mosses, once lakes. They have been mapped and described by Mr. Geikie. They also occur in Northumberland.

for I have shown that in rocky regions, the farther
north you go, the more do lakes increase in number,
and I am convinced that this fact is not a mere accidental coincidence, but is one of the strongest proofs
of the former existence of that wide-spread coating of
glacier-ice, that in old times moulded the face of so
much of the northern hemisphere.

This theory, brought out in March, and published in
August 1862, of the glacial origin of so many lakes in
the northern hemisphere, or rather wherever there have
been either continental or mountain glaciers, was on
the whole received with disfavour or 'faint praise' in
England when first produced, and it fared no better in
Switzerland, and but little better in the north of Italy,
where however it was allowed that it 'deserved the
gravest attention.' Nevertheless, it begins to make
way among some of those physical geologists whose
opinions I highly value. Mr. Geikie in his 'Phenomena
of the Glacial Drift of Scotland,' has declared that it
will be 'at last accepted as the true theóry of the
origin of those rock-basins of the Alps and of the
northern hemisphere generally that are occupied by
lakes.' Sir William Logan, in his report on the geology
of Canada for 1863, has drawn attention to it in a foot-
note, and in the text he says, that the great North
American lake-basins 'are depressions not of geological
structure but of denudation; and the grooves on the

surfaces of the rocks which descend under their waters, appear to point to glacial action as one of the great causes which have produced these depressions.' Dr. J. S. Newberry, in the American 'Annual of Scientific Discovery' for 1863, pp. 252 and 255, has also drawn special attention to the subject; and Dr. Otto Torrel, who has given so much attention to the glacial geology of Greenland, Spitzbergen, and Scandinavia, informed me that my theory threw an unexpected light on the Swedish lakes.*

* Since the text was printed, Professor H. Y. Hind of Toronto has informed me that in 1855–6 he read a paper before the Canadian Institute of Toronto, 'On the Origin of the Basins of the great American Lakes,' in which he attributed that origin to ice, but the memoir was never printed. He has also pointed out to me the following passage in his work on the 'Canadian Red River, and Assinniboine, and Saskatchewan Expeditions,' published in 1860: 'The wide-spread phenomena exhibiting the greater or less action of ice, such as grooved, polished, and embossed rocks, the excavation of the deep lakes of the St. Lawrence basin, the forced arrangement of drift, the ploughing up of large areas, and the extraordinary amount of the denudation at different levels without the evidence of beaches, all point to the action of glacial ice previous to the operations of floating ice in the grand phenomena of the drift.'

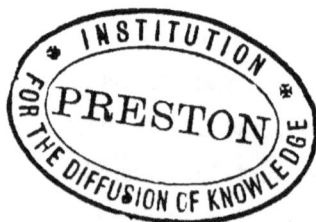

LECTURE V.

NEWER PLIOCENE EPOCH, CONTINUED. BONE-CAVES.
DENUDATION OF THE COASTS OF BRITAIN. BRITISH
CLIMATES AND THEIR CAUSES; AREAS OF DRAINAGE,
RIVER VALLEYS, AND THEIR ORIGIN; OLD RIVER GRA-
VELS, AND PRE-HISTORIC HUMAN REMAINS. HISTORICAL
ELEVATION OF THE COUNTRY.

I HAVE already said that during the younger Tertiary
epochs, England has been repeatedly joined to the
main-land, a circumstance proved by the remains of
terrestial mammalia, the bones of which are found im-
bedded in the strata. Thus, to take one late instance,
it is clear that England must have been united to the
Continent before the formation of the Mammaliferous
Crag (one of the Newer Pliocene deposits), because we
find in these beds the remains of a number of terrestrial
quadrupeds, many of them of great size. Later still, at
Cromer, in Norfolk, there is a bed known as '*the
forest-bed,*' because there are a number of trees im-

bedded in it. In this 'forest-bed' are found the bones of the *Rhinoceros Etruscus*, the *Elephas meridionalis*, and the great Mammoth, known to naturalists as the *Elephas primigenius*—according to Dr. Falconer the oldest strata in which this elephant has yet been found. Such large mammalia, on any hypothesis, did not all originate in a small detached island like England, but formed parts of large families of Pachydermata that inhabited the north of Europe, America, and Asia at various periods of geological time, and they could only have passed into our area by the union of England with the Continent.

Again, in the south of England, at Selsey Bill, there are post-pliocene strata on the sea-shore described by Mr. Godwin-Austen, one of the beds containing species of living marine shells, not belonging to icy seas, and overlaid by icy boulder-drift, and in the former there were found the remains of another well-known species of elephant, the *E. antiquus*, lying on clay on which stumps of trees, the remains of an old wood, still stand.

These boulder-drifts were formed during a period of cold, accompanied by a vast extension of the great glaciers that covered so much of the north of Europe and America, as I have already explained. While the boulder-drift was forming, the country slowly sunk, in our country so far that the mountains of Wales,

Cumberland, the higher parts of Derbyshire and of Scotland, became merely groups of small islands. But the cold continued so intense, even during that period of submergence, that these islands were surrounded by ice, and more or less covered by perpetual snow. Then, of course, our country was entirely severed from the mainland; but after this period it was again elevated, and there is evidence that it was again united to the Continent, for we now find the remains of a number of new species of animals unknown in the older sub-formations. The Mammoths which lived before this time must have been driven out of our area by that submergence, unless it be possible that a few, with other mammalia, managed to live on in the extreme south of what is now England, which apparently only suffered a very small change of level. Farther north such large animals could not have lived on mere groups of icy islands, on which, if there was any vegetation, it was exceedingly scanty. Such animals required a large amount of vegetation to feed them, and it is therefore clear that they must have died out or been banished from our area by that submergence. We find, however, that on the re-elevation of the country, it must have been re-united to the Continent, because this great hairy Elephant again appears, and is associated with a number of other animals that also migrated from the Continent of Europe to our area. Among these there is the

Rhinoceros tichorhinus, and various oxen, one of which,
·the Auroch, is still living in the forests of Lithuania.
We have also, in the old lake and river-beds of that
period, Hippopotami, Horses, the great Deer known as
the *Cervus megaceros,* commonly called the Irish Elk;
the Reindeer; the wild Ass; a large Bear known as
the *Ursus spelæus;* a great Tiger; the Leopard; the
Lynx; the wild Cat of the existing species; the *Hyæna
spelæa,* an extinct species, along with wolves, foxes,
otters, beavers, and a number of other small animals.
Besides these, found in alluvial, gravelly, and old lake
deposits, their remains are also found in the bone-
caves of the country. These bone-caves are often of
very old date, and always occur in limestone strata, in
which they have been formed in consequence of part
of the lime being dissolved. Most of the solid limestone rocks are exceedingly jointed, and rain-water
finding its way down and in among the strata through
the joints, the carbonic acid in the water by degrees
dissolves part of the limestone in the form of bicarbonate of lime, and running in underground channels
sometimes as large rivers, caves have been formed often
of great extent, and branching in many directions, and
in these caves the remains of extinct mammals are
found mingled with the bones of species that still in-
habit our country and the Continent of Europe. Sometimes they seem to have been washed in through the

mouths of the caverns, sometimes they fell or were washed in through openings in the roof; and not un-' frequently the detached bones of animals, or the animals themselves, have been dragged in by beasts of prey, such as the old Hyena that inhabited these caves. One evidence of this is, that the bones are frequently gnawed, for very commonly you may detect on them the marks of the teeth of carnivora. Another proof of these caves having been inhabited (as shown by Dr. Buckland) is, that the sides of the caverns themselves are occasionally smooth, having been polished by the animals rubbing against the rock as they passed by corners and along other uneven surfaces in their way in and out of the recesses of their dens.

There is no doubt that many of these caves date from before the epoch of the icy drift, and also that the bones of animals found their way into some of them before that period; and since it closed, many of them have been more or less tenanted as caverns down to the present day, or bones have been at intervals washed into them, and thus it happens that organic remains perhaps of older date than the drift are found in the same cave with bones belonging to the icy period, and to minor epochs of still later date. Mingled with the bones of extinct species in various parts of England and Wales, flint implements and other remains of man have been found, and though it is usually said that

they are always of later date than the boulder-beds,
it is by no means certainly proved that this is the case.
Some of the Devonshire caves in which the works of
man were found were above water during the drift
period, and others farther north, like that of Cefn in
North Wales, were below the sea, for the boulder beds
reach a higher level; and along with Dr. Falconer I
found fragments of marine shells of the drift, in the
cave overlying the detritus that held the bones. No
human remains were found in that cavern, and it is just
possible that some of those in the south of England
may have been above water and inhabited, while others
farther north lay underneath the icy sea.

After the elevation of the country that succeeded
the drift period, the probabilities are that England was
united to the Continent, not by a mass of solid rock
above the sea level, but by a continuation of the
boulder-drift over what is now the German ocean; and
across this plain many of the animals of which I have
spoken migrated into our area, some of them for the
second time. It is the belief of some geologists that at
the same period Ireland was united to England and
Scotland by a similar plain across the area now covered
by the Irish sea, and over this the *Megaceros Hiber-
nicus*, formerly called the Irish Elk, and a number of
other animals migrated. The proof is equally clear
that Ireland during the drift period, like England, was

in great part submerged, so as to form a group of small islands, and therefore to allow of the country being re-inhabited by large mammals, there must have been ground over which these mammals walked into the Irish area, after the re-elevation of the country.

An excellent surmise was offered us on this subject by the late Professor Edward Forbes, who drew attention to some remarkable observations made by the late Mr. Thompson of Belfast, with regard to the comparative number of reptiles that are found in Belgium, in England, and in Ireland. In Belgium there are in all 22 species of serpents, frogs, toads, lizards, and the like. In England the number of species is only 11, and in Ireland 5, and the inference that Professor Forbes drew was, that these reptiles migrated from east to west, across the old land that joined our island to the Continent, before the denudations took place that disunited them. Before the breaking up of that land a certain number of them had got as far as England, and a smaller number as far as Ireland, and the continuity of the land being broken up, their further progress was stopped.

These denudations, of course, did not cease with the breaking up of the land that joined our territory to the Continent, and there are clear proofs of several oscillations of the relative levels of sea and land since that period. This waste of territory is indeed going

on still, and will always go on while a fragment of
Britain remains. Before proceeding further, I would
advance one or two proofs, to show you how steady the
waste of our country is, that you may know what I
have said to be founded on observation.

Along the east coast of England, between Flam-
borough Head and Kilnsea, the strata are composed of
what I have called drift or boulder-clay, sometimes
of great thickness, and forming well-marked sea-cliffs.
This district is called Holderness, and many towns,
built upon the coast, have been forced by degrees to
migrate landwards, because of the encroachment of the
sea. 'The materials,' says Professor Phillips, 'which fall
from the wasting cliff,' (a distance of 36 miles) 'are
sorted by the tide, the whole shore is in motion, every
cliff is hastening to its fall, the parishes are contracted,
the churches wasted away.' The whole area on which
Ravenspur stood, once an important town in Yorkshire, where Bolingbroke, afterwards Henry IV., landed
in 1399, is now fairly out at sea. The same may be
said of many another town and farmstead, and the
sea is ever muddy with the wasting of the unsolid land.
In like manner, all the soft coast cliffs, from the Humber to the mouth of the Thames, are suffering similar
destruction, in places at an average rate of from 2 to 4
yards a year. One notable example is found at Eccles-
by-the-sea in Norfolk. The town at a comparatively

late period extended beyond the church tower, which
is now buried in blown sand, and the church itself has
been destroyed.

, On the south side of the estuary of the Thames,
stands the ruined church of the Reculvers on a low
hill of Thanet sand, half surrounded on the land side
by the relics of a Roman wall that in old times en-
circled the little town. The church has been abandoned, but is preserved as a landmark by the Admiralty,
and groins have been run out across the beach to prevent the further waste of the cliff by the sea. As it
is, all the seaward side of the Roman wall has long
been destroyed, the waves have invaded the hill, and
half the church-yard is gone, while from the cliff the
bones of men protrude, and here and there lie upon
the beach. A little nearer Herne Bay the same marine
denudation sparingly strews the beach with yet older
remains of man, in the shape of flint weapons of a
most ancient type, washed from the gravels that crown
part of the cliff. In the Isle of Sheppey, great slips
are of frequent occurrence from the high cliff of London
clay that overlooks the sea.*

Again, in the Tertiary basin on the south coast of
England, if you walk along the foot-paths that are used

* Two acres of wheat and potatoes have in this manner slipped seaward in 1863. When I saw them the crops were still standing on the
shattered ground below the edge of the cliff.

by the coast-guardsmen, you often find that the path on the edge of the cliff comes suddenly to an end, and has been re-made inland. This is due to the fact, that the cliffs, chiefly composed of clay and sands, are so soft, that, as in Sheppey and Holderness, every year large masses of country slip out seaward and are rapidly washed away.

The waste of this southern part of England and of Holderness, has been estimated at the rate of from two to three yards every year. You will easily see, therefore, that in the course of time, a great area of country must have been destroyed. At Selsey Bill, there is a farmhouse standing about 200 yards from the shore, and the farmer told me, that since he first settled there, as much land has been wasted away as that which now lies between his house and the sea; and the site of the Saxon Cathedral Church that preceded that of Chichester, is said to be far out at sea. But this waste is not confined to the softer kinds of strata, for further west in Dorsetshire, where Oolites and Chalk form the cliffs, we find the same kind of destruction going on, one remarkable case of which is the great land-slip in the neighbourhood of Axmouth, which took place in the year 1839. The strata there consist on the surface of chalk, underlaid by greensand, which is underlaid by the Lias clay. The chalk is easily penetrated by water, and so is the sand that

underlies it. After heavy rains the water sinking through
the porous beds, the clay beneath became exceedingly slippery, and thus it happened that the strata
dipping seaward at a low angle, a vast mass of chalk
nearly a mile in length slipped out seaward, forming a
grand ruin, the features of which are still constantly
changing, by the further foundering of the chalk and
greensand. The waves beating upon the foundering
masses destroy them day by day, and in time they will
entirely disappear. If you walk along that coast and
criticise it with a geological eye, you will observe that
a great number of similar land-slips have taken place
in times past, of which we have no special record.

In parts of our country further west, the Silurian
rocks, Old red sandstone and Coal-measures on the
coast, show evidence of waste; as for instance at St.
Bride's Bay in Pembrokeshire, where the north and
south headlands are formed of hard igneous rocks that
stand boldly out seaward; while between these points
there are softer Coal-measure strata, which undoubtedly
once filled what is now the bay—and probably with the
other strata spread far beyond. But because of their
comparative softness they have been less able than the
igneous rocks of the headlands to stand the wear and
tear of the atmosphere and the sea waves, and thus
they have been worn back, and a large bay is the
result. Indeed, all along the west coast where solid

rocks prevail, hard rocks usually form the promontories, while the bays have been scooped in softer material; and this, though the rate of waste may not be detected in many years, yet proves its nature. The very existence of sea cliffs proves marine denudation. I merely mention these things to show that denudations on a great scale are going on now, and therefore when I speak of former unions and separations of our island with the main land by denudation, and oscillation of level, the statement is founded on excellent data.

I now come to other phenomena connected with the physical structure of our island, and its geography generally; and first, with regard to the rain that falls upon its surface. If you examine the best hydrographic map of the Atlantic, you will see numerous lines and arrows showing the direction of the flow of the ocean currents as drawn by Captain Maury. One great current begins in the Gulf of Mexico, where the water in that land-locked area within the tropics is exceedingly heated; and flowing out of the gulf it first passes E. and then NE., across the Atlantic and so reaches the European area of the North Sea. So marked is the heat of this immense current, that in crossing from England to America, the temperature of the water suddenly falls some degrees. Several years ago, in crossing the Atlantic, I was in the habit early in the morning of taking the temperature of the water with one of the officers of the steam-boat. I then found that at about five o'clock in the morning, for several days, the temperature of the sea was always about 4 degrees above the temperature of the air, but quite suddenly, in passing out of the Gulf stream, at the same hour of the morning the temperature of the water was found to average about 4 degrees below the temperature of the air. Where in the map the arrows point southwards, there are cold streams of water coming from the icy seas of the north. One of these passes along the east coast of America, and coming from the north sea and floating many an iceberg, the western half of the north Atlantic is thus kept cool, and the water is often colder than the air.

Now the Gulf stream occupies a very great width in the Atlantic, and approaches tolerably near to our own western coast, and the effect of this great body of warm water flowing northward, is to divert the isothermal lines (lines of equal temperature) far to the north, over a large part of the Atlantic area. Thus a certain line runs across North America, about latitude 50°, representing an average temperature for the whole year of 32°. Across that continent it passes tolerably straight, but no sooner does it get well into the Atlantic, than the Gulf stream flowing north-wards, warms the air, and the result is, that the line bends away to the far north above Norway; thus in the west of Europe producing an average warmer climate, of the whole year, than exists in corresponding latitudes in North America, the middle of Europe, and the interior of Asia. Our British climate and all the west of Europe, thus becomes as it were abnormally warm, owing to the influence of the Gulf stream, and you will at once recognise this fact, from the circumstance that trees of goodly size grow much further north on the west coasts of Europe than on the east coasts of North America. Another effect that the Gulf stream produces, is to cause an unusual amount of moisture in the west of Europe, and if you consult a rain map of the British islands you will see represented by different shades the average amount of rainfall in different areas—the darker the shade the greater the quantity of rain. Now the prevalent winds in the west of Europe are from the SW., and therefore during a great part of the year the south-west wind comes laden with moisture across the land, warm and moist from the sea where the Gulf stream flows.

In the extreme south-west of England, in Cornwall, about thirty-five inches of rain fall every year; and in the western parts of the north of Scotland, and roughly speaking, over great part of Ireland, the amount of rain-fall is about thirty inches; while for much of the middle and east of England we have a fall of about twenty-five inches. The climate, therefore, of

Great Britain is varied in the fall of rain, and the average temperature of the western area is also raised and rendered agreeable by the influence of the Gulf stream. So much is this the case, that certain garden plants grow through the winter in Wales and the West of England, and even in the far north-west of Scotland, which the winter cold of Middlesex kills. Now this moisture in the air is partly intercepted on its passage eastward by the mountains which rise in the west of the British islands. Every one who has visited Cumberland and Wales knows what rainy regions they are, compared with the centre and east of England, because the moisture that is blown from the Atlantic is precipitated upon the mountains and adjacent lands. The same is the case in Scotland, where the Highland mountains that rise on the west produce the same effect in intercepting the moisture, and thus, partly because it is the first land that the wind laden with moisture reaches, and partly because of the mountains, it happens that a greater amount of rain is precipitated in the western than in the eastern parts of our island.

Now, if we examine our country with regard to special areas of drainage, we find that they are exceedingly numerous. The Spey, which runs into the German ocean, drains an area of very nearly 1,200 square miles. The Tay drains an area formed by the Grampian mountains, and part of the Old red sandstone of 2,250 square miles. The Forth, including its estuary, drains an area of about 2,000 square miles. The Clyde, not including the greater part of its estuary, drains an area of 1,580 square miles.

If we take the Trent and the Ouse as draining one area, the immense extent, for such a country as ours, of about 9,550 square miles are drained into the Humber. The Thames drains an area of about 6,000 square miles; and if we include all the estuary, about 10,000. The Severn drains an area of 8,580 square miles; and many others of importance might be mentioned, too numerous to name here.

Now, it is difficult even approximately to settle what are the geological dates of the valleys through which the rivers run, or, in other words, when they first began to be scooped out, and through what various periods their excavation was intermittently or continuously carried on. No one has yet analysed this subject, and, for my part, I only begin to see my way into it. Nevertheless a little may be done even now, and a great deal will be accomplished when with sufficient data the whole subject may come to be investigated. In Wales, for example, there are vast numbers of rivers and brooks, small and large, and when you examine the relation of these streams to the present surface of the country you often find it very remarkable. Fig. 14 (Map, line 14) is a diagram repre-

Fig. 14.

senting no particular section, but simply the nature of the real sections across the Lower Silurian strata of Cardiganshire, as shown by myself in a paper given to the British Association at Oxford in 1847. The dark-coloured part represents the form of the country, given in the original sections on a true scale of six inches to a mile horizontally and vertically. The strata of this area, and, indeed, of much of South Wales, are exceedingly contorted. The level of the sea is represented by the lower line, and if you take a straight-edge, and place it on the topmost part of the highest hill, and incline it gently seaward, it touches the top of each hill in succession in the manner shown by the line *b b*. This line is as near as can be straight, and on the average has an inclination of from one to one-and-a-half degrees; and it is a curious circumstance that in the original lines of section, there were no peaks rising above that line, —they barely touched it as in the diagram, and no more. It occurred to me, when I first observed this circumstance, that, at a period of geological history of unknown date perhaps older than the beginning of the

New red sandstone, *this inclined line that touches the hill-tops must have represented a great plain of marine denudation.*

The sea waves on the cliffs by the shore are the only power I know that can denude a country, *so as to shave it across* and make a plain either horizontally or slightly inclined. If a country be sinking very gradually, and the rate of waste by the waves on the shore be proportionate to the rate of sinking, a little reflection will show you that the result would be a great inclined plane like that of the straight line *b b* in the diagram. Let South Wales be such a country : then when that country was again raised out of the water, the streams made by its drainage immediately began to scoop out valleys ; and though some of the valleys may have been begun by marine denudation during the process of emergence, yet in the main I believe that the inequalities below the line *b b* have been made by the influence of running water. Hence the number of deep valleys, many of them steep-sided, that diversify Wales, all the way from the Towey in Caermarthenshire to the southern flanks of Cader Idris and the Arans. On ascending to the upper heights, indeed, anywhere between the Vale of Towey and Cardigan Bay, it is impossible not to be struck with the average uniformity of elevation of the flat-topped hills that form a principal feature of the country. Between the rivers

Towey and Teifi, and in other areas, these hills in fact form the relics of a great plain or table-land *in which the valleys have been scooped out;* and in the case of the country represented in fig. 14, 'the higher land, as it now exists, is only the relics of an average general gentle slope represented by the straight line (*b b*) drawn from the inland heights towards the sea.' *   My colleague, Mr. Jukes, has lately applied and extended the scope of the same kind of reasoning to the south of Ireland, with great success.   In various parts of Europe, notably in those regions that have been longest above the water—on the banks of the Moselle and of the Rhine—you find innumerable valleys, intersecting table-lands of a form that leads you to believe that they have been made by the long-continued action of running waters, and I believe that the South Wales valleys have been formed in a great measure in the same way.

But when we come to the region of the larger rivers, there is a great deal that is difficult to account for. One thing is certain, that before the glacial epoch, which I described with so much detail in the last lecture, the greater contours of the country were much the same as they are now.   The mountains of Scotland, Wales, and of Cumberland, and the great Pennine chain, existed then as now; the central plains of England were plains then, and the escarpments of the Chalk and Oolites

* Reports, British Association, p. 66, 1847.

existed before the glacial period. All that the ice did was to modify the surface by degradation, to smooth its asperities by rounding and polishing them, to deepen valleys where glaciers flowed, and to scatter quantities of detritus in the shape of boulder-clay and sands and gravels, over the plains that form the east of England, and the Lias and New red sandstone in the middle. If we examine the valley of the Severn from Bristol northwards through Coalbrook Dale, we find that for a large part of its course the river runs down a great valley between the old palæozoic hills and the escarpments formed by the table-land of the Cotswold range, that rises so high in the neighbourhood of Cheltenham. That valley certainly existed before the glacial epoch, because we find boulders and boulder drift far down towards Tewkesbury, and therefore I believe, that before the glacial epoch this part of the Severn ran very much in the same course that it does at present. Then the country sank beneath the sea, and Plinlimmon itself, where the river rises, was buried in part, or possibly altogether, beneath the waters. When the country again emerged, the old system of river-drainage in that area was resumed, and the Severn following in the main its old course, cut a channel for itself through the boulder-clay that partially blocked up the original valley in which it ran. But *when* that original valley was formed through which the older Severn ran, no

man can yet say, although England having probably
been above the sea during great part, or perhaps the
whole of the undoubted Miocene epoch, it is likely
that some of our greater contours were then first begun,
or if not begun, carried on, and very seriously modified.

Again if we take the rivers Forth and Clyde; as I
explained to you in a previous lecture, the areas occupied by these rivers and by the rocks through which
they run, is an exceedingly ancient area of depression.
That country is also covered more or less with boulder
clay, and with later stratified detritus of sand and
gravel which were formed in part by the remodelling of
the glacial drifts. The great rivers I believe ran in
that area before the commencement of these deposits,
and for aught I know to the contrary for very long
before that period. But we have no perfectly distinct
traces of these earlier epochs in that part of Scotland,
and all we know is that rivers ran in these valleys
before the deposition of the boulder clay, and resumed
to some extent their old courses after the emergence
of the country. Again, if we examine the channels of
other rivers in the east of England, as for instance the
Ouse in Bedfordshire—we find that some of them flow
through areas covered with this clay, and have cut
themselves channels through it in such a way as to
lead to the inference that the valleys in which they run
did not exist before the boulder-bed period, but that

they have cut their courses through it and the underlying Oolitic strata, and formed a new system of valleys.
These, however, are exceptional, and often only apply
to parts of their channels.

Again, with regard to the Thames, it is remarkable
that it rises in the Seven Springs not far from the edge
of the Oolitic escarpment of the Cotswold hills that
overlook the Severn, which runs in the valley about
1,000 feet below. The infant Thames thus flows at
first across a broad table-land of Oolite, and by and
by comes to a second table-land, formed of the Chalk,
and the wonder is, that there its course was not turned
aside by that high escarpment. Instead of that being
the case, a valley cuts right across the escarpment of
chalk, through which the river flows. This escarpment is exceedingly old—dating from long before the
boulder-beds, but how long no one knows, for we find
far-transported boulders in places at its base, while in
the same neighbourhood the drift has not been deposited on its slopes nor yet does it lie on the top. How
did the Thames find its way through what was once
that great unbroken scarped barrier of chalk now called
the Chiltern Hills? Not that such phenomena are confined to this river alone—it is a trick that rivers have;
they *will* cut through escarpments in what seems an
unnatural fashion. Mr. Jukes, in a paper read before
the Geological Society in June 1862, on the River-

L

valleys of the South of Ireland, has attempted an explanation of this, with regard to those rivers, and I firmly believe that his solution holds good in many cases in England, as for instance in those that cut across the escarpments of the Weald mentioned at page 82. In the south of Ireland, Mr. Jukes finds that some rivers break suddenly right across chains of hills, and his explanation of the circumstance is that when a certain river began to flow, the surface of the country formed a kind of inclined plain, in the manner shown at p. 140, part of that old surface being then the top of the range of hills across which the river now flows; and just in proportion as it cut a channel downwards so as to form the present valley *at the foot of, and along the strike of the hills, so without needful fracture it also deepened by degrees its channel across the range.* It is astounding when first thought about, because the time required to perform such an operation is so stupendous that the mind almost refuses to grapple with it. But the more we ponder on geological time, the more is required for every individual subject; and I for one easily grant it for the channel-cutting power of rivers.

I will show you, however, that this process may not necessarily always be required to cut a passage through a range such as the Chalk escarpment. If you follow the Oolitic strata northwards, you will see that they pass into the channel of the Humber, and reappearing on the opposite side, pass north into Yorkshire. The Chalk escarpment does the same, and both lie low. Now by the side of the Humber, both Chalk and Oolites are at the level of the sea, and it is possible that the sea beating upon the rocks, during various old oscillations of level, before the drift period, may in the course of time have cut a channel for itself across the ranges of low Chalk and Oolitic hills in that area now partly occupied by the estuary of the Humber. If this be true, the sea effected a breach in the rocks. Then suppose these lands to have been heaved up, so that the present salt-water estuary became high and dry; the river then ran through it, and we have a river-valley cutting right across the escarpment. It seems not impossible, that at an old period of the history of the estuary of the Thames, the escarpment of the Chalk through which the river now flows may at a lower. level have been breached by the sea in the way I have supposed with the estuary of the Humber. Then, the whole being heaved up, the Thames flowed through the gap in the Chiltern hills, and across the London clay down to the present sea. Though worth consideration, I do not however depend on this hypothesis, for as I have no doubt that the gaps both of the Humber and the Thames are of far older date than the glacial period, their precise origin is in the present state of the subject lost in the mists of geological antiquity. It may be that Mr. Jukes' hypothesis may apply even to the Thames, the Ouse, and the Humber, the origin of their scarp-cutting valleys dating from a time inconceivably remote in years, when both Cretaceous and Oolitic rocks spread much further west than they do now, having with a low eastern dip in some old period been simply cut by denudations into an inclined plane sloping east, in a manner analogous to that of the Weald already described.

On the banks of the Thames and many other rivers there are frequent terraces. It is one of the effects produced by the action of rivers to form terraces upon their banks; close to or at a distance from the river as it now is, according to its size and other circumstances. The hills on either side are, perhaps, made of solid rock, or of softer strata, as the case may be, and the terraces lying between the higher slopes and the rivers consist of gravel, sometimes of old date, remodelled by these rivers as they cut their way from side to side, thus by degrees deepening their valleys. A river for instance at one time flowed over the top of the highest gravel terrace, and winding about from side to side of the valley and cutting away detritus in its course, it formed various terraces one after another, the terrace on the highest level being of oldest date, and that on the lowest level that bounds the modern alluvium the latest. Thus in the following figure, Nos. 1 and 2 represent the solid rocks of a country bounding 'a wide valley partly filled with ancient gravel, No. 3, which originally filled the valley from side to side as

Fig. 15.

high as the uppermost dotted line 4, but a river flowing through by degrees bore part of the loose detritus to a lower level, thus cutting out the terraces in succession marked No. 5, 6, and 7. Or again, in other cases (as in the Moselle for example), where no special valley of the present shape existed before the drainage took the general direction of its present flow, the river has actually excavated its own valley by the destruction of the solid rocks through which it flows. In the occasional terraces which accompany processes such as this, it often happens that alluvial and gravelly deposits are left marking ancient levels of the rivers. Or it sometimes happens that when the physical geography of the country was a little different from what it is now, such an ancient valley was in places partly filled by irregular strata deposited by the river itself (as in the Ouse near Bedford), through which in time, as circumstances changed, the river cut its way, and formed terraces as in fig. 15. It is in the gravel, sand, and loam of such terraces that the works of old races of men have often been found along with the bones of extinct animals.

In the year 1847, a French *savant*, Mons. Boucher de Perthes, of Abbeville, published an account of these remains in the first volume of his 'Antiquités Celtiques.' In examining the old gravels of the river Somme, he discovered the remains of terrestrial animals, some of them of extinct species. The strata consisted of surface soil, below which was nearly five feet of brown clay, then loam, then a little gravel containing land shells, and along with these shells the teeth of the Mammoth. Below that, there occurred white sand and fresh-water shells, and again the bones and teeth of the Mammoth and other extinct species; and along with these bones and teeth, a number of well-formed flint hatchets. Geologists were for long asleep on this subject. Mons. de Perthes had printed it many years, but none of them paid much attention to him. At length, Mr. Prestwich having his attention drawn to the subject, began to examine the question. He visited Mons. de Perthes, who distinctly proved to him and afterwards to other English geologists that what he had stated was incontestably the fact. These hatchets are somewhat rude in form, but when I say 'rude,' I do not mean that there is any doubt of their having been formed artificially. They are not polished and finished, like those brought from the South Sea Islands; but there can be no doubt whatever that they have been formed by the hand of man; and I say this with authority, since, for more than twenty-five years, I have been daily in the habit of handling stones, and no man who knows how chalk flints are fractured by nature, would doubt the artificial character of these ancient tools or weapons.

Of late the same kind of observations have been made in our own country. In the neighbourhood of Bedford—on the Ouse, there are beds of river gravel of this kind which rise about 25 feet above the level of the river—lying in broad terraces; and in one of these, far above the level of the river, there have been found associated with river shells, the bones of the Mammoth, old varieties of oxen, and various other mammalia, and along with these a considerable number of flint hatchets. By the river Waveney also, on the borders of Norfolk and Suffolk, near Diss, the same phenomena have been observed in old gravel pits, made for the extraction of road materials; and it has been proved that near the mouth of the estuary of the Thames between the Reculvers and Herne Bay, flint hatchets have fallen from the top of a cliff of Eocene sand capped with recent gravel. These were first noticed by Mr. T. Leech, and I have myself found one of them partly water-worn by the waves. No bones have as yet

been observed in that district along with the implements. But it is very clear that the bones of *Elephas*
primigenius and of other extinct mammalia occur
in many places associated with the works of pre-
historic men. As yet, however, the bones of men
have never in our country been discovered along with
them in the gravels. In certain British caves in
limestone rocks—such as Kent's hole near Torquay,
Brixham cave in the same county, those in Gower,
Glamorganshire, and in several others, remains of the
Mammoth, two species of Rhinoceros, together with the
Cave bear, Hyæna, and many other animals of species
both extinct and still living have been found, and with
these not only flint knives, flakes, and hatchets were
discovered, but implements made of bone, such as
needles or hair pins, and various other articles. On
the continent of Europe, also, in caverns in the limestone on the river Meuse in Belgium, Dr. Schmerling
found bones of men and other living species mingled
with those of extinct mammals, such as the Cave bear,
Hyæna, Elephant, and Rhinoceros. His account, long
slighted, though full of errors of detail, has at length
obtained the attention it deserved. Further, within the
last few years, in the surface strata called the Loess of
the same river near Maestricht, human skeletons are said
to have been actually found. I have seen these remains,
which certainly have an antique look, but some doubt

exists as to their authenticity. In the same neighbourhood, however, it is certain that a human jaw was
found in strata containing the remains of Mammoths,
&c. Many other examples might be given of the remains of old races of men in such like caverns or in
river deposits, but enough has been said to show that
there can be no doubt that man was contemporary in
our island and elsewhere with extinct mammalia, and
it is possible that his origin in our island dates as far
back as the time when the country was united to the
mainland, and that along with the great hairy Mammoth,
the Hippopotamus, the Irish Elk, and the Rhinoceros he
travelled across, at a time when the arts were so rude
that he had no other means of coming *except upon his
feet.*

One word more on a kindred subject, and then I
shall close this lecture. Round great part of our coast
we find terraces, from twenty to fifty feet above the
level of the sea, and in some places the terrace runs
with great persistence for a number of miles. Round
the Firth of Forth, for example, on both shores, there
is an old sea-cliff of solid rock, overlooking a raised
beach or terrace, now often cultivated in cornfields and
meadows, and then you come to the present sea beach.
This terrace usually consists of gravel and sea-shells in
great quantities, of the same species with those that lie
upon the present beach, where the tide rises and falls

every day. The same kind of terrace is found on the shores of the Firth of Clyde, and in almost all the other estuaries of Scotland, and in places round the west highlands on the coast of Scotland. Similar or analogous raised beaches occur on the borders of Wales, and in the south of England. In Devon and Cornwall, there are the remains of old consolidated beaches clinging to the cliffs from twenty to thirty feet above the level of the sea. It is clear, therefore, that an elevation of the land has occurred in places to the extent of about forty feet, at a very recent period, long after all the living species of shell-fish inhabited our shores. Further, in the alluvial plains that border the Forth, and on the Clyde in the neighbourhood of Glasgow, at various times, in cutting trenches, canals, and other works, the bones of whales, seals, and porpoises have been found at a height of from twenty to thirty feet above the level of high water-mark. Now it is evident that whales did not crawl twenty or thirty feet above high water-mark to die, and therefore they must either have died upon the spot where their skeletons were found or been floated there after death. That part of the country therefore must have been covered with salt water, which is now occupied simply by common alluvial detritus. But the story does not stop there, for in the very same beds in which the remains of these marine mammalia have been discovered on the Clyde,

canoes have been found in a state of preservation so
perfect, that all their form and structure could be
well made out. Some of them were simply scooped in
the trunks of large trees, but others were built of
planks nailed together,—square-sterned boats indeed,
built of well dressed planks,—and the inference has
been drawn by my colleague, Mr. Geikie, who has described them, that this last elevation took place at a
time that is historical, and even since the Roman occupation of our island.

There is one piece of evidence with respect to the
very recent elevation of these terraces which I think is
deserving of great attention, and it is this:—In the
neighbourhood of Falkirk, on the south shore of the
Firth of Forth, there is a small stream, and several
miles up that stream, beyond the influence of the tide
of the present day, there were at the end of last
century remains of old Roman docks, near the end of
the Roman wall usually called the wall of Antoninus
that stretched across Scotland from the Firth of Clyde
to the Firth of Forth. Those docks are now no longer
to be seen : but so perfect were they, that General Roy
was able to describe them in detail, and actually to
draw plans of them. When they were first built they
were of course close to the tide, and stood on the
banks of a stream called the Carron, believed by
Mr. Geikie to have been tidal; but the sea does not

come near to them now ; and therefore he naturally inferred, that when they were constructed the relative height of the land to the sea must have been less than at present. Again, the great wall of Antoninus, erected to keep out the northern barbarians from the territory conquered by the Roman legions, must have been brought down close to the sea level at both ends. Its eastern termination is recognised by most antiquarians as having been placed at Carriden, on the top of a considerable cliff, where the great Falkirk flats disappear along the shore. Its western extremity, not having the favourable foundation of a steep rising ground, now stands a little way back from the sea-margin of the Clyde. When it was built it was probably carried to the point where the chain of the Kilpatrick Hills, descending abruptly into the water, saved any further need for fortification. But owing to a probable rise of the land, a level space of ground twenty or twenty-five feet above the sea now intervenes between high water-mark and the base of the hills and runs westward from the termination of the wall for several miles as far as Dumbarton. Had this belt of land existed then, there appears little reason to doubt that the Romans would not have been slow to take advantage of it, so as completely to prevent the Caledonians from crossing the narrow parts of the river, and drive them into the opener reaches of the estuary below Dumbarton.

LECTURE VI.

QUALITIES OF WATERS. CONNECTION OF THE PHYSICAL
GEOLOGY OF THE COUNTRY WITH THE POPULATION.

IN last lecture I gave a sketch of the chief river
areas of Great Britain, but I did not enter upon
one important point connected with them, namely, the
qualities of their waters. If we examine the geological
structure of our island, with regard to its water-sheds
and river courses, we find that the larger streams, with
one or two exceptions, run into the German Ocean, the
chief exception being the Severn and its tributaries,
which drain a large proportion of Wales, and a considerable part of the interior of England. The reason
why the larger rivers run chiefly to the east is, that in
consequence of the nature of old disturbances and
denudations of the rocks, the main water-shed of the
country from the north of Scotland far into England
is nearer to the western coast than to the eastern,
and thus it is that a much larger area of country is

drained towards the east than to the west. On the west side of England we have more than one tract of high ground, and in the middle and east, plains and table-lands of which I had occasion to speak in a former lecture, and the area occupied by these flats being much larger than the western hilly districts, the rivers there are larger, with, as I have already said, the exception of the Severn.

Now, when we look at the qualities of the waters of rivers, we find that this depends on the nature of the rocks and soils over which they flow. Thus the waters of the rivers of Scotland are, for the most part, soft. All the Highland waters, as a rule, are soft; the mountains being composed of granitic rocks, gneiss, mica-schist and the like, a very small proportion of limestone being intermingled therewith, and the other rocks being comparatively free from lime. Only a comparatively small proportion of lime, soda, or potash, is taken up by the water that falls upon, flows over, or drains through these rocks, the soda or potash being chiefly derived from the feldspathic ingredients of the various formations, and therefore the waters are soft. For this reason, a few years ago, at a vast expense, Glasgow was supplied with water derived from Loch Katrine, which, lying amid the gneissic rocks, is like almost all other waters from our oldest formations, peculiarly soft, pure and delightful. The same is the case with the waters that run from the Silurian rocks of the Lammermuir hills, except where these are covered by a drift of foreign materials.

The water from the Welsh mountains is also in great part soft, the country being composed of Silurian rocks, here and there slightly calcareous from the presence of fossils mixed with the hardened sandy or slaty sediment, that forms the larger part of that country. From this cause so sweet and pleasant are the waters of Bala lake, compared with the impure mixtures we drink in London, that it has been more than once proposed to lead it all the way for the supply of water for the capital. But when in Wales and on its borders we come to the Old red sandstone district, the marls are somewhat calcareous, and interstratified with impure concretionary limestones, called cornstones, and the waters are hard. The waters are apt to be still harder in the Carboniferous limestone tracts that sometimes rise into high and almost mountainous ridges round the borders of the great South Wales coal-field, and in Flintshire and Denbighshire. Again, the waters which flow from the Pennine chain, that extends from the southern borders of Scotland into Derbyshire, are all hard, because they drain areas composed chiefly of Carboniferous limestone; and all the rivers that run east from this range, and all those that flow in areas as. far south as the British Channel, over the New red sandstone and Lias, and the Oolitic and the Cretaceous rocks, are of necessity charged with those substances in solution that make water hard.

Before proceeding to other subjects I should like to give you some idea of the quantity of certain salts which are carried in solution into the sea by the agency of running water.

The first case I shall take is at Bath, where there is a striking example of what a mere spring can do. The Bath Old well yields 126 gallons of water per minute, which is equal to 181,440 gallons per day. There are a number of constituents in this water, such as carbonate of lime, nearly nine grains to the gallon; sulphate of lime, more than eighty grains to the gallon; sulphate of soda, more than seventeen grains to the gallon; common salt, rather more than twelve and a half grains to the gallon; chloride of magnesium, fourteen and a half grains to the gallon; etc. etc.— altogether, with other minor constituents, there are 144 grains of salts in solution in every gallon of this water, which is equal to 3,402 lbs. per day, or 420 tons a year. Now a cubic yard of limestone may be roughly estimated to weigh one ton. If, therefore, these salts were precipitated, compressed, and solidified into the same bulk as limestone, we should find the annual discharge of the Bath wells to form a square column 9 feet in diameter and about 140 feet high. Yet this large amount of solid mineral matter is carried away every year in invisible solution in water which, to the eye, appears perfectly limpid and pure.

Again, the Thames is a good type of what may be done in this way by a moderate-sized river, draining a country which to a great extent is composed of calcareous rocks. It rises in the middle of England at the Seven Springs, near the western edge, and therefore not far from the top of the great Oolitic table-land, of the Cotswold hills, and flows eastward through all the Oolitic strata, which are composed mostly of thick formations of limestone, calcareous sand, and masses of clay, which often contain shelly bands and scattered fossil shells. Then, bending to the south-east, below Oxford, it crosses the Lower greensand, the Gault, the Upper greensand, and the Chalk, the last of which may be roughly stated as consisting of nearly pure limestone ; then through the London clay, and other strata belonging to the great Eocene formation of the London basin, which are nearly all more or less calcareous. The Thames may therefore be expected to contain numerous substances of various kinds in solution in large quantities; and to those derived from the rocks must be added all the impurities from the drainage of the villages and towns that line its banks between the Seven Springs and London.

At Teddington, on a rough average, 1,337 cubic feet

of water (equal to 8,329 gallons) pass seaward per second ; and upon analysis it was found that twenty-two and a half grains of various matters, chiefly bicarbonate of lime, occur in solution in each gallon, thus giving 187,402 grains per second passing seaward. This is equal to 87,844 lbs. per hour, or 33,497 tons per annum, and this amount is almost entirely dissolved *out of the bulk of the solid rocks and surface soils of the country,* and is passing out to sea in an invisible form, and only known to the analytical chemist. If you consider that this is only one of many rivers that flow over rocks which contain lime and other substances easily soluble, you will then see what an enormous quantity of matter by this—to the eye—perfectly imperceptible process is being gradually carried into the sea. And it is a necessary part of the economy of nature that it should be so, for it is from salts thus obtained by the sea that plants and shell-fish derive, to some extent, their nourishment.

This waste of material by the *dissolving* of rocks is indeed evident to the practised eye over most of the solid limestone districts of England, and I shall therefore say a little more on the subject. On the flat tops of the Chalk Downs, for example, over large areas in Dorsetshire, Hampshire, and Wiltshire, quantities of angular unworn flints, many feet in thickness, completely cover the surface of the land, revealing to the

thoughtful mind the fact that all these thick accumulations of barren stones have not been transported from
a distance, but rather represent the gradual destruction
by rain and carbonic acid of a vast thickness of chalk
with layers of flint that once existed above the present
surface. The following diagram will explain this.

Fig. 16.

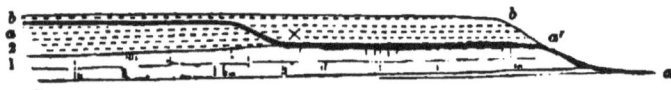

1, Chalk without flints. 2, Chalk with flints. *a a*, the present surface
of the ground marked by a dark line. *b b*, an old surface of ground,
marked by a light line. Between *a a* the surface is covered by accumulated flints, the thickness of which is greatest where the line is
thickest between *a'* and × , above which surface a greater proportion of
chalk has been dissolved and disappeared.

There can be little doubt but that the great plateaus
of Carboniferous limestone and Oolite have suffered
waste by solution, equal to that of the Chalk, only from
the absence in them of flints—we have no insoluble
residue by which to estimate its amount.

The soils of a country necessarily vary to a great
extent, though not entirely, with the nature of the underlying geological formations. Thus, in the Highlands
of Scotland the gneissic and granitic mountains are
generally heathy and barren, because their hard rocky
materials frequently come bare to the surface over great
areas. Strips of more fertile meadow land lie chiefly

on narrow alluvial plains, which here and there border
the rivers. Hence the Highlands mainly form a wild
and pastoral country, sacred to grouse, black cattle, sheep
and red deer. Further south similar rocks, though the
scenery is different, yet produce more or less the same
kind of soil, in the broad range of hills that lies
between the great valleys of the Clyde and Forth,
and the borders of England, including the Moorfoot
and the Lammermuir hills, and the high grounds that
stretch southwards into Carrick and Galloway. The
rocks there, being composed of hard untractable gritty
and slaty material, form but little soil, because they
are difficult to decompose. Hence the ground being
mostly high, is to a great extent untilled, though ex-
cellently adapted for pastoral purposes. Where, however, the slopes descend, and are covered more or less
with old ice drifts and moraine matter, the soil is
deeper and the ground is more fertile.

The great central valley of Scotland, between the
metamorphic rocks of the Highland mountains and the
less altered Silurian strata of the high-lying southern
counties, is occupied by rocks of a more mixed character, consisting of Old red sandstone and marl, and of
the shales, sandstones and limestones of the Coal-measures intermixed with considerable masses of igneous
rocks. The effect of denudation upon these various
rocks in old times, particularly of the denudation which

took place during the glacial period, and also and more especially of the rearrangement of the ice-borne débris by subsequent marine action, as the country sunk beneath the sea and rose again, has been to cover large tracts of country with a happy mixture of materials— such as clay, mixed with pebbles, sand and lime. In this way one of the most fertile tracts anywhere to be found in our island has been formed, and its cultivation for nearly a century has been taken in hand by skilful farmers who have brought agriculture over great part of that district up to the very highest pitch which it has ever attained in any part of Great Britain.

Through the inland parts of England from Northumberland to Derbyshire, we have another long tract of hilly country, composed of palæozoic rocks, forming in part such high regions that much of it is unfitted for ordinary agricultural operations. A considerable part of it is therefore devoted to pasture land, as is also the case with large portions of Cumberland and the other north-western counties of England. The same features are observable in Wales, where disturbance of the palæozoic rocks has resulted in the elevation of a great range, or rather of a cluster of mountains—the highest south of the Tweed. In that old Principality and in the Longmynd of Shropshire there are great tracts of land, amounting to thousands upon thousands of acres, where the country rises to a height of from

1,000 to 3,500 feet above the level of the sea. Much
of it is covered with heath and is therefore fit for
nothing but pasture land; but on the low grounds
and on the alluvium of the rivers, there is often excellent soil. When we come to the eastern part of this
hill-country in Monmouthshire, Brecknockshire, Herefordshire, and parts of Worcestershire, occupied by the
Old red sandstone, the larger proportion of the country
—though hilly, and in South Wales occasionally even
mountainous—is naturally of a more fertile kind, from
the circumstance that the rocks are much softer and
therefore more easily decomposed; and where the
surface is covered with drift, the loose material is
chiefly formed of the waste of the strata on which it
rests, and this adds to its fertility. The soil is thus
deepened and more easily fitted for purposes of tillage.
In the centre of England, in the Lickey hills, near
Birmingham, and in the wider boss of Charnwood
Forest, where the old palæozoic rocks crop out like
islands amid the Secondary strata, it is curious to
observe that a wild character suddenly prevails in the
scenery, for the rocks are rough and untractable, and
stand out in miniature mountains. Much of Charnwood Forest is, however, covered by drift, and is now
being so rapidly enclosed, that, were it not for the
modern monastery and the cowled monks who till
the soil, it has almost ceased to be suggestive of the

England of mediæval times, when wastes and forests covered half the land.

If now we pass to the Secondary rocks that lie in the plains, we find a very different state of things. In the centre of England formed of New red sandstone and marl, the soils are for the most part more fertile than in the mountain regions of Cumberland and Wales, or in some of the Palæozoic areas in the extreme south-west of England. When the soft red sandstone and marl are bare of drift, and form the actual surface, they often decompose easily, and form deep loams, save where the conglomerate beds, lying in the middle of the New red sandstone, come to the surface. These conglomerates consist to a great extent of gravels barely consolidated, formed of well waterworn rounded pebbles, of various kinds, but chiefly of liver-coloured quartz-rock derived from some unknown region, and of sili-ceous sand, sometimes ferruginous. This mixture forms therefore, to a great extent, an exceedingly barren soil. Some of the old waste and forest lands of England, such as Sherwood Forest and Trentham Park, lie almost entirely upon these intractable gravels, or on other barren sands of the New red sandstone, and have partly remained uncultivated to this day. As land however becomes in itself more valuable, the ancient forests are being cut down and the ground enclosed. But a good observer will infer from the straightness and smallness of the hedges, that such ground has only been lately taken into cultivation, and at a time since it has become profitable to reclaim that which at no distant date was devoted to forest ground and to wild animals.

In the centre of England there are broad tracts of heavy land composed chiefly of New red marl and Lias clay. When you stand, as I stated in a previous lecture, on the summit of the great escarpment, formed by the Oolitic table-land, you look over the wide flats and undulations formed of this New red marl and Lias clay. The marl consists of what was once a light kind of clay, mingled with a small per-centage of lime, and when on the surface it again moulders down, it naturally forms a fertile soil. A great extent of the arable land in the centre and west of England is formed of these red strata; and it is worthy of notice that the fruit-tree districts of Great Britain lie chiefly upon red rocks, sometimes of the Old and sometimes of the New red series. The counties of Devonshire, Herefordshire, and Gloucestershire, with their numerous orchards, celebrated for cider and perry, lie in great part on these formations, where all the fields and hedgerows are in spring white with the blossoms of innumerable fruit trees. Again, in Scotland the plain, called the Carse of Gowrie, lying between the Sidlaw Hills and the Firth of Tay, stretches over a tract of Old red sandstone, and is famous for its apples.

What may be the reason of this relation I do not know; but such is the fact, that soils composed of the New and Old red marl and sandstone are better adapted for fruit trees than any other in Britain.

The Lias clay in the centre of England forms a considerable proportion of our meadow land. It is blue when unweathered, and includes many beds of limestone, and bands of fossil shells scattered throughout the clay itself. From its exceeding stiffness and persistent retention of moisture, it is especially adapted for grass land, for it is not easy to plough, and thus a large proportion of it in the centre of England is devoted to pastures intersected by innumerable foot-paths of ancient date, that lead by the pleasant hedge-rows to wooded villages and old timbered farms. When we pass into the Middle Lias, which forms an escarpment overlooking the Lower Lias clay, we find a very fertile soil; for the Marlstone, as it is called, is much lighter in character than the more clayey Lower Lias, being formed of a mixture of clay and sand with a considerable proportion of lime, derived from its numerous fossils. The course of the low flat-topped Marlstone hills striking along the country and overlooking the Lower Lias clay, is thus usually marked by a strip of peculiarly fertile soil, often dotted with villages and towns with antique churches and towers built of the brown limestone of the formation.

When we ascend into the next geological group, we find the Oolitic Downs, formed, for the most part, of beds of limestone, with here and there interstratified clays, some of which, like the Oxford and Kimeridge clays, are of great thickness, and spread over considerable tracts of country. The flat tops of these limestone Downs, when they rise to a considerable height, as they do on the Cotswold hills, were until a very recent date left in a state of natural grass, and were used chiefly as pasture land. They formed a feeding ground for vast numbers of sheep, but they are being brought by degrees under the dominion of the plough, and on the highest of them you now find fields of turnips and grain. The broad flat belts of Oxford and Kimeridge clay that lie between the western part of the Oolite and the base of the chalk escarpment are still in great part used as pastures for cattle.

If we pass next into the Cretaceous series, which in the middle and south of England forms extensive tracts of country, we meet with many kinds of soil, some, as those on the Lower Greensand, being excessively siliceous, and, in places, intermingled with a marked proportion of oxide of iron. Such a soil still remains in many places, intractable and barren. Thus on the western borders of the Weald towards Petersfield, where there is very little lime in the rocks, there are wide-spread heaths on which you may walk for hours

over ground unenclosed and almost as wild and refresh-
ing to the smoke-dried denizens of London as the
broad moors of Wales and the Highlands of Scotland.
These, partly from their height, but chiefly from the
poverty of the soil, have never been worth bringing
into a state of cultivation. Beneath these beds of Greensand lies the Weald clay, which is now almost entirely
cultivated and improved by help of deep drainage.
Below this Weald clay, various members of the Hastings
sand crop out, forming extensive tracts of country, all
belonging to the Wealden series, and forming in the
main the undulating hills that lie half-way between
the North and South Downs. The sand of these beds
is for the most part so fine that when dry it may be
described as an almost impalpable siliceous dust, and
within their area lay, and still lie in part, the old uncultivated forests of Ashdown, Tilgate, and St. Leonards.
Down to a comparatively late historical period, both
clays and sands were left in their native state, forming
those broad forests and furze-clad heaths that covered
almost the whole of the Wealden area. Hence the
name Weald or Wold (a woodland), a Saxon term,
applied to this part of England, though the word does
not now suggest its original meaning, unless to those
who happen to know something of German, or Saxon
derivatives.

The Chalk strata of the South Downs stretch far

into the centre and west of England. South of the
valley of the Thames they form the North Downs, and
stretch north-east into Yorkshire in a broad unbroken
band. Most Londoners are familiar with these Downs,
lying, as they do, so near. ˙In their wildest native state
where the ground lies high these districts were probably
almost bare of woods— 'the bushless downs'— and they
are still largely used for pasturage, yet here, also, cultivation is gradually encroaching. Broad sweeping plains,
even Salisbury Plain itself, which, within my own recol-
lection, were almost entirely devoted to sheep, are
being gradually invaded by the plough, and turned into
arable land. Many of the slopes of the great chalk
escarpments on the North and South Downs, in the
West of England, on the Chiltern hills and elsewhere,
are however so steep, that the ground, covered with
short turf, and in places dotted with yew and juniper, is
likely to remain for long unscarred by the ploughshare.

The clay bands of the Eocene beds, occur on all
sides of London. They are often covered by superficial
sand and gravel. Through the influence of the great
population centred here, originally owing to facilities
for inland communication afforded by the river; this is
now, in great part, a highly cultivated territory. Here
and there, however, to the south-west, there are tracts
forming the lower part of the higher Eocene strata,
known as the Bagshot sands, which produce a soil

so barren that, although not far from the metropolis, it is only in scattered patches that they have been brought within the influence of cultivation. They are still for the most part bare heaths, and being sandy or dry, we place camps upon them, and use them as exercise-grounds for our soldiers. Higher still in this Eocene series lie the fresh-water beds on which the New Forest stands, commonly supposed to have been depopulated by the Conqueror and turned into a hunting ground. But to the eye of the geologist it easily appears that the wet and unkindly soil produced by the clays and gravels of the district form a sufficient reason why most of the hundred villages, said to have been destroyed, perhaps never existed there, for the soil for the most part is barren, and probably grew a great native forest even in the Conqueror's day.

Such is a very imperfect sketch of the general nature of the soils of Great Britain, and of their relation to the underlying rocks. We have seen that throughout large areas, the character of the soil is directly and powerfully influenced by that of the rock-masses lying below. It must be borne in mind, however, that the abrading agencies of the glacial period have done a great deal towards commingling the detritus of the different geological formations, and thus producing a wide-spread drift of varied composition. This drift is, however, far from being uniformly spread over the island. In some districts it is absent, while in others it forms a thick mantle, obscuring all the hard rocks and giving rise to a soil sometimes nearly identical with that produced by the waste of the underlying formation, and sometimes of mixed clay and stones as in Holderness. Thus it is often poor and often of the most fertile description, as in certain upper members of the formation in the great valley of the Lothians.

I shall now say a few words on the influence of the geology upon the inhabitants of different parts of our island.

Great Britain is inhabited by several great races, more or less intermingled with one another. It requires but a cursory examination to see that the more mountainous and barren districts, as a whole, are inhabited by two Celtic populations, very distinct from each other, and yet akin, while the lowland parts are occupied by the mixed descendants of various other races.

From the earliest historical period it appears that both sides of the English Channel were inhabited by a Celtic people, known to us by the name of the Cimri (Cymru),—ancient Britons, whom we now call Welsh. Further north another Celtic people, the Gael, inhabited the greater part of what is now termed Scotland, the Isle of Man, and Ireland. Analyses of modern Welsh and Gaelic prove, in the opinion of some accomplished scholars, that these Celtic branches, now so distinct, yet sprung from the same original stock. Nevertheless I believe that the Gael as a people is much more ancient in our islands than the Cymru ; and I think it may be proved, almost to demonstration, that the ancestors of the Scottish Highlanders (who however are now largely intermixed with Scandinavian blood) once spread not only much farther south than the borders of the Highlands, but that they even occupied the Lowlands of Great Britain generally, for the names of many of the rivers in England and even in Wales have a Gaelic and not a Welsh origin, complete or in combination. Thus, all the rivers called Ouse, the Usk, the Esk, the Don, and others, derive their names from the Gaelic. It is a characteristic of rivers often to retain the names given them by a primitive race long after that race has been expelled, and thus the Gaelic *Uisge* (water) has not in all cases been replaced by the ancient Welsh *Gwy.* This old Welsh word we constantly find in a corrupt form, as in the Wye, the Medway, the Tawe, the Towey, and the Teifi, the Dovey and the Dove ; or the water of the rivers is expressed in another form by the later *dwfr* or *dwr* as in Stour, Aberdour, &c. In both languages, river (*Avon*) is the same. If the earlier inhabitants were Gaelic, then they were driven northwards into their mountains, by the superior power of another and later Celtic population that found its way to our shores, and pushed onwards, occupying the more fertile districts of England and the south of Scotland; for the Gael would not willingly have confined themselves to the barren mountains if they could have retained a position on more fertile lands.

Thus I believe it happens that the north of Scotland, beyond the great valley, is chiefly inhabited by the Gael. On the east, however, along the coasts of the Moray Firth, in Caithness and in the Orkney and Shetland Islands, the people are of Scandinavian origin and speak Scotch, thus standing out in marked contrast with the Gaelic clans, who possess the wilder and higher grounds in the interior and western districts. There is here a curious relation of the human population to the geological character of the country. The Scandinavian element is strongly developed along the maritime tracts, which, being chiefly composed of Old red sandstone, stretch away in long and fertile lowlands; while the Celts are pretty closely restricted to the higher and bleaker regions where the barren gneissic and schistose rocks prevail. It is remarkable that a number of the names of places in the centre and south of Scotland are not Gaelic, but have been given by the later race and can be translated by anyone who has even a superficial knowledge of Welsh, and it is almost certain that from the Lowlands of Scotland all through the midland and southern parts of Britain, the country was inhabited in the later Celtic times by the same race that now peoples Cornwall and Wales. The names of scores of places, now unintelligible to the vulgar, proves it. Thus there are all the Coombs (*Cwm*) of Devon and Somersetshire, and even as far east as Croydon; Dover (*dwfr*); the Cumbraes (*Cymru*), in the Clyde, and Cumberland; and at Bath, by the Avon, we have 'Dolly (*dolau*) meadows;' near Birmingham 'the Lickey hills' (*llechau*); near Macclesfield the rocky ridge called 'the Cerridge' (*cerrig*); and in the hills of Derbyshire 'Bull gap,' the Welsh *bwlch*, translated, just as in another instance *dolau* is repeated in the English word meadows. Again in the lowlands of Scotland we have *Aberdour* (the mouth of the water), Lanark (*Llanerch*, an open place in a forest, or clearing), Blantyre (*Blaen tir*, a promontory or projecting land), Pennycuik (*Pen-y-gwig*, the head of the thicket), and many other corrupted Welsh names. The wide area over which this language was spoken is indeed proved by the ancient Welsh literature, for the old heroic poem of the Gododin was composed by Aneurin, said to have been a native of Strath Clwyd in Scotland. But however this may be, it is certain that the British Celts, when the Romans invaded our country, overspread the whole of the southern part of Great Britain. By and by, they mixed with their conquerors, but the Romans, as

N

far as blood is concerned, seem to have played an
unimportant part in our country. They may have
intermarried to some extent with the natives, but
they occupied our country very much in the manner that we now occupy India. Coming as military
colonists, they mostly went away as soon as their
time of service was up, and finally abandoned the
country altogether. But after the retirement of the
Romans, invasions took place by the Scandinavian
tribes, the Anglo-Saxons, and others who came to
occupy the land permanently. Then the native
tribes, dispossessed of their territories and driven westwards, retreated into the distant and mountainous parts
of the country, where the relics of this old Celtic
people are still extant in part of Devon and in Cornwall,
while among the Welsh mountains the same Celtic
element still forms a distinct and peculiar people.
There, till after the Norman conquest, they could still
hold out against the invader, and maintain their inde-
pendence in a region barren in the high ground, but
traversed by many a broad and pleasant valley.
Living, as the relics of the old Britons are apt to do,
so much in memories of the past, the slowly dying
language, and even the antique cadences of their re-
gretful music, speak of a people whose distinctive characters are waning and merging into a newer phase of
intellectual life.

It appears then that the oldest tribes now inhabiting

our country, both in Scotland and in the south, are to
be found among those most ancient of our geological
formations, the Silurian rocks, which, by old palæozoic
disturbance; form the high mountain lands, while the
lower and more fertile hills, the plains and table-lands,
are chiefly inhabited by the descendants of 'the heathen
of the northern sea,' who made good their places by the
sword after the departure of the Romans.

To enter in detail upon the peculiar effect of geology
on the industry of the various races or the populations
of different districts, would lead me far beyond the
proposed scope of these lectures. I shall, therefore,
only give a mere outline of what ought to be a course
of lectures, rather than by any means attempt to
exhaust the subject.

First, let us turn to the older rocks. In Wales these
consist, as I have already stated, to a great extent of
slaty material. The largest slate quarries in the world
lie in the Cambrian rocks of Carnarvonshire. One
single quarry, that of Penrhyn, is half a mile in
length, and more than a quarter of a mile from side to
side. A number of other quarries occur in the same
district, but none of them of the same vast size. Some
other important quarries also lie in the Lower
Silurian rocks near Ffestiniog in Merionethshire. In
these districts there is a large population which is

chiefly supported by the quarrying and manufacture of slates. The slate quarry at Penrhyn, near Bangor, presents a wonderful spectacle of industry. More than 3,000 men are there employed solely in the making of slates, which are exported to all parts of the world, and yield a vast revenue to the proprietor, the Hon. Colonel Douglas Pennant. Other quarries at Llanberis, which belonged to the late Mr. Assheton Smith, employ nearly an equal number of men in comfort, and the revenues to the proprietor are proportionate. There are many other smaller quarries in the neighbourhood, while further south in Merionethshire some are worked in caverns instead of open day. So great is the profit derived from slate quarries that every here and there in Wales, where the rocks are more or less cleaved, speculators go to work and opening part of a hill-side, find a quantity of rotten stuff, or of slate full of iron pyrites, or cut up by small joints, or imperfectly cleaved, and after a time being ruined, they sell the property to other speculators, who ruin themselves in turn. There are slate quarries in South Wales, Cumberland, and in some parts of Scotland, but they are all unimportant compared with the immense quarries of North Wales.

In various districts of Great Britain the rocks abound in the ores of certain metals, and, occurring in part in hilly pastoral regions, the workers in these mines are rarely congregated in great crowds like the slate quarriers of North Wales, or the miners of coal and iron. I will first allude to the cases in which the mineral wealth is derived from what are termed lodes, from whence are derived our chief supplies of copper, tin, zinc, and lead. It is worthy of remark that these lodes are almost wholly confined to our oldest or palæozoic rocks. The Devonian formation contains them in Devon and Cornwall, and the Silurian formations in Wales, Cumberland, in the hills of the south of Scotland, and here and there throughout the Highlands. In the Carboniferous limestone they are also found in North Wales, Cumberland, and Derbyshire. The chief districts in England where copper and tin are found are in Devon and Cornwall; and in Wales, especially in Cardiganshire and Montgomeryshire, there are ores of copper, and many lodes highly productive in ores of lead, some of which are rich in silver. No tin mines occur in that district. Gold also has of late years been found in Merionethshire between Dolgelli, Barmouth, and Ffestiniog, sometimes as at Clogau, in profitable quantity, but generally only in sufficient amount to form pretexts for getting up companies which sometimes lure unwary speculators to their loss. This Welsh gold is found in lodes near the base of the Lingula flags, which in that area are talcose, and pierced by eruptive bosses of igneous rocks and greenstone dykes.

In older times extensive gold mines were worked
in Caermarthenshire at the Gogofau (*ogofau*, caves),
between Llandovery and Lampeter. These excavations
were first made open to the day, in numerous irregular
shallow caverns where the gold-bearing quartz veins
were followed into the hill; and later by means of
lofty well-made galleries, which cut the lodes deeper
underneath. The gold was also found in washings of
the superficial gravel of nearly a mile in length on the
banks of the river Cothy. The galleries and the washings are clearly Roman, but it has been surmised by
the proprietor Mr. Johnes that the ruder caverns partly
date from more ancient British times. The huge excavations must have made ugly scars on the hills in the
day when they were worked, but time has healed them.
The heaps of rubbish are now green knolls, and gnarled
oaks and ivy mantle the old quarryings.

In the Carboniferous limestone districts, both of
North Wales and Derbyshire, there are numerous lead
mines; and, as I have already said, it occurs in the
underlying Silurian strata, as in South Wales and also
in the Lead Hills in the south of Scotland, where lead
associated with silver, and even a little gold, has long
been worked. There are numerous veins of lead and
other ores in the limestone of the Mendip Hills.

I must now endeavour to give you an idea of what
a lode is. A lode is simply a crack, more or less

filled with various kinds of mineral matter, such as layers and nests of quartz, carbonate of lime, carbonate of copper, sulphuret of copper, sulphuret of lead, oxide of tin, or with other kinds of ore. Various theories have been formed to account for the presence of ores in these cracks, and to this day the subject is not perfectly clear. Formerly, the favourite hypothesis was, that they were formed by a sublimation from below, somehow or other connected with the internal heat of the earth ; and the ores were supposed to have been deposited in the cracks through which the heated vapours passed. A great deal also has been said on the effect of electric currents passing through the rocks, and aiding in depositing along the sides of fissures the minerals which were being carried up by sublimation from below, or were in solution in waters that found their way into the fissures. I dare not utter any positive statement on the question, but my opinion is that the ores of metals in lodes have generally been deposited from solutions. We know that water, especially when warm, can take up silica in solution and deposit it, as in the case of the Geysers in Iceland ; and we also know that metals may, in some states, be held in solution in water, both warm and cold. This is proved by the accurate results of chemists, who have detected silver and gold, and, it is said, copper, in solution in sea water. We must remember that, when the lodes or cracks were originally formed, those parts of them that we explore were not so near the surface as we now see them; but in a great many, and perhaps in all cases, they lay deep beneath the surface, covered by thousands of feet of rock that have since been removed by denudation. They were therefore probably, in all cases, channels for subterranean waters, both in their upper portions that have been removed by denudation and in the originally deeper parts that now remain. It is not unlikely also that these subterranean waters may often have been warm, seeing that they sometimes lay deep in the interior of the earth and came within the influence of the so-called central heat, whatever may be its origin. For my own part I do not doubt that the ores which we meet with in these cracks or lodes were formed by infiltration, for strings of copper, lead, and tin, for example, occur in the mass just in the same way that we find mixed with them strings of carbonate of lime or quartz. If this be so, then, just as the lime and silica may have been derived from the percolation of water through the rocks that form the country on each side of the lode, so the metalliferous deposits seem to have been derived from metalliferous matter minutely disseminated through the neighbouring formations. We are, however, still in the dark as to many of the conditions under which the process was carried on.

Ores of iron are common in lodes and in hollows or

pockets, both in the limestones of the Devonian and Carboniferous periods.

In the Coal-measures, however, we have our greatest sources of mineral wealth, because they have been the means of developing other kinds of industry besides that which immediately arises from the discovery of the minerals which the Coal-measures contain. In the great coal-fields of this formation occur all the beds of coal worth working in Britain. In the South Wales coal-field there are more than 100 beds of coal, great and small, and one celebrated bed in South Staffordshire attains a thickness of 40 feet. By a fortunate geological coincidence, from the very same Carboniferous strata—for nearly 100 years past—we have obtained almost all the iron that has been employed in our manufactures. Of late years, however, a great deal of valuable iron ore has been obtained from the Marlstone of Yorkshire, and the Northampton sands which are the geological equivalents of the Stonesfield slate. In older times, in the Weald of the south of England, a considerable amount of iron used to be mined and smelted with wood or charcoal, before the Coal-measures were worked extensively, and when the Weald was covered to a great extent with forest. Then the chief part of our iron manufactures was carried on in the south-east of England. Indeed, late in the last century, there were still iron furnaces in the Weald of Kent and Sussex.

The last furnace is said to have been at Ashburnham, and every here and there you may now see heaps of slags overgrown with grass, and the old dams which supplied the water that drove the water-wheels that worked the forges of Kent and Sussex. It is said that the cannon that were used in the fight with the Spanish Armada came from this district; and the rails round St. Paul's were also forged from the Wealden iron.

Besides coal and iron, the Coal-measures yield quantities of clays, which are of considerable value. The chief of these is fire-clay, which is used so largely in the manufacture of crucibles, fire-bricks, and furnaces. But by far the most important mineral which this geological formation affords is 'coal. Instead of occurring in veins and lodes, as many of the metals do, it is spread out in vast sheets, or layers, bed above bed, within the crust of the earth. If you look at the geological map of England, you will see that large patches are coloured black—they are the Coal-measure districts of Great Britain. Some of these coal-fields, as, for instance, the great coal-field of South Wales and the Forest of Dean, lie in a basin-shaped form, and the coal crops to the surface all round the basin. But in other parts of England the coal does not occur in basins, but crops to the surface at one side alone, the remainder being shrouded by deep coverings of New red sandstone, and of other Secondary rocks. Thus it sometimes happens that only by a series of geological accidents have the Coal-measures been brought to the surface, exposed to view, and rendered available for use. We may take the South Staffordshire coal-field as an example, where the New red sandstone and Permian rocks are thrown down against the coal-field on both sides. Originally, before these faults took place, the New red sandstone and other rocks spread entirely over the surface. The New red sandstone and marl, where thickest, are more than 2,000 feet thick; above it lies the Lias, 900 to 1,500 feet thick; then come the Oolites, and lastly all the Cretaceous strata. This enormous mass of superincumbent strata once lying above the south Staffordshire coal-measures was afterwards dislocated by faults, which brought the lower portions of them down against the sides of the present coal-field. A vast denudation must have ensued, whereby a great thickness of the formations nearest the surface was removed, and the whole country was worn down to one comparatively general level. It is by such processes that some of our large and productive coal-fields have been exposed at the surface. Hence we now find a great manufacturing population all centred in areas (like those of South Staffordshire, Warwickshire, and Ashby-de-la-Zouch), which might never have been known to contain coal-fields, had it not been for the geological accidents of those faults and denudations which I have repeatedly

noticed and explained. In other cases, as in the great
South Wales coal-field, that of the Forest of Dean,
and the coal-field of the great central valley of Scotland
through which the Clyde and the Forth run, the Carboniferous strata were probably never covered by Secondary strata. Between the mouth of the Firth of
Clyde and the mouth of the Firth of Forth the whole
country is one great coal-field, and this is the part of
Scotland where the population is thickest. Bordering
Wales and the mountains of Lancashire and Derbyshire,
on the east and west, are three great coal-fields, and
these districts also contain dense populations. Further
north lies the great Newcastle coal-field, where, again,
the population is proportionately redundant. All the
central part of England, which is dotted over with coal-
fields, teems in like manner with inhabitants. The
South Wales coal-field, which is the largest of all, however, does not, except in places such as Swansea and
Llanelly, show the same concentration of population.
This exception is due to peculiar circumstances. A
great part of this area being exceedingly hilly, the coal
has been heretofore by no means worked to the same
extent as in the coal-fields of the middle and northern
parts of England, which have been extensively mined
for a longer period.

I have already remarked that a large part of the
wealth which we owe to our Carboniferous minerals,

arises, not so much from the commercial value of these minerals themselves, as from the fact that they form the means of working many different branches of our industry. To the vast power which steam has given us, very much of our extraordinary prosperity as a nation is due. Yet were it not for our coal-beds, the agency of steam would be almost wholly denied to us. And hence it is that our great manufacturing districts have sprung up in the vicinity of coal-fields. There iron furnaces glare and blow day and night, there are carried on vast manufactures in all kinds of metal, and there our textile fabrics are chiefly made. In these busy scenes a large part of the population of our island finds employment, and thence we send to the farthest parts of the earth those endless commodities, which, while they have supplied the wants of other countries, have given rise in large measure to the wealth and commerce of our own.

There are some other geological formations which afford materials for manufactures. Thus, in the south-west of England, in the granitic districts of Devon and Cornwall, and a little further east—in Dorsetshire, near Poole—a great proportion of the finer kinds of clays occur, which are used in making stoneware and porcelain. In Devon and Cornwall the decomposition of granite affords the substance known by the name of Kaolin, from which all the finer porcelain clays of our

country are made. It is formed by the disintegration
of the felspar of granite. This felspar, as I showed in
a previous lecture, consists of silicates of alumina, and
soda or potash. The soda and potash are comparatively
easily dissolved, chiefly through the influence of carbonic acid in the water that falls upon the surface; and
the result is that the granite decomposes to a considerable depth. In some cases I have seen granite
which had never been disturbed by the hand of man,
which for a depth of twenty feet or more might be
easily dug out with a shovel. Owing to this decomposition, a portion of the felspar passes into kaolin,
which is washed down by the rains into the lower levels,
where, more or less mixed with quartz, and the other
ingredients of granite, it forms natural beds of comparatively pure clay. This is dug out, and the clay is
transported chiefly to the district of the Potteries in
North Staffordshire. The same process is sometimes
secured by art, when the decomposed granite being dug
out, is washed by artificial processes, and the more
aluminous matter is separated from the quartz with
which it was originally associated. Then, in the Potteries, it is turned into all sorts of vessels — fine porcelain, stone-ware, and common-ware, in every variety
of size, and form, and texture.

In the Eocene tertiary beds in the neighbourhood of

Poole, there are large lenticular beds of pipe-clay, interstratified with the Bagshot sand. Great quantities
of this clay are exported into the Pottery districts to be
made into the coarser kind of earthenware, and they
are also mixed with the finer materials from Devon and
Cornwall, to make intermediate qualities of stone-ware
and china.

But in addition to clay, the chalk is brought into
requisition to furnish its quota of material for this
manufacture. The flints that are found embedded in
the chalk, chiefly in layers, are also transported to the
Potteries, and ground up with the aluminous portions
of the clay, since it is necessary to use a certain proportion of silica in the manufacture of porcelain.

Many other formations, such as the Old and New
red marls, are also of use when clay is required. The
Oolitic and Liassic strata are to a great extent composed of clay, such as Lias clay, Fuller's earth, Oxford
and Kimeridge clay, and the Gault lies in the middle
of the Cretaceous strata. An abundance of material is
found in these formations for the manufacture of bricks,
earthenware pipes, and so on; and it is interesting to
observe how in this respect the architecture of the
country is apt to vary according to the nature of the
strata of given areas. In Scotland and the north of
England, where hewable stone abounds, almost all the

houses are built of sandstone, grey and sombre; in
many of the Oolitic districts they are of limestone, and
lighter and more graceful; while on the Lias and in
the Woodland area of the Weald we have still the relics
of an elder England in those brick and timbered houses
that speak of habits and manners gone by.

In the Lias strata, chiefly in the upper Lias clay
in Yorkshire, beds of lignite and jet are found near
Whitby, which, however, do not form anything like
an important branch of manufacture.

The glass-sand used in this country is chiefly derived
from the Eocene beds of the Isle of Wight, and from
the sand-dunes on the borders of the Bristol Channel.
In the Isle of Wight, the sandy strata lie above the
London clay, and are the equivalent of part of the
Bagshot sands.   They are remarkably pure in quality,
being formed of fine siliceous white sand.   These sands
are largely dug and exported to be used in glass-houses
in various parts of the country, as in Birmingham and
elsewhere.

A large proportion of the cement stones of our
country comes from the Lias limestone.   These limestones are not pure carbonate of lime, but are formed
of an intermixture of carbonate of lime and aluminous
matter.   It is found by experience that the lime from
this kind of limestone is peculiarly adapted for setting
under water. Hence the Lias limestone has always been

largely employed in the building of piers and other structures that require to be constructed under water. Cement stones are also found to some extent in the Eocene strata, and are obtained from nodules dredged from the sea-bottom at Harwich, and the south of England. These are transported hither and thither, to be used as occasion may require.

The chief building stones of our country, of a hewable kind, are the limestones of the Oolitic rocks, the Magnesian limestone, the Carboniferous limestone, and the Carboniferous sandstones. The chief Oolitic building stones are from the Isle of Portland and the Bath Oolite. St. Paul's was built of Portland stone, and the mmense quantities of rejected stones in the old quarries, show how careful Sir Christopher Wren was in the selection of material. The Carboniferous sandstones in Lancashire, and in the neighbourhood of Leeds, Edinburgh, and Glasgow afford a large quantity of admirable building material, which has been used almost exclusively in the building of these towns. Some of it is exceedingly white, it is easily cut by the chisel, and may be obtained in blocks of immense size. But in some of the beds there is so much diffused iron, not visible at first sight, that in the course of time this, as it oxidises, forms dark stains which discolour the exterior of the buildings. The New red sandstone also yields its share of building stones, but much of it is very soft and easily worn by the weather, a notable example of which may be seen in the cathedral of Chester. The white Keuper sandstone of Grinshill, north of Shrewsbury and elsewhere, is an excellent stone. The Old red sandstone is also used as a building stone in its own area, and the Caradoc sandstone of Shropshire yields a beautiful white material.

In Devonshire and Cornwall, and in Scotland, but chiefly near Aberdeen, the granite quarries afford much occupation to a number of people. Now that it has become the fashion to polish granites, these rocks are becoming of still more importance. But as they are not so easily hewed as sandstone, they do not come into use as ordinary building stones, except in such districts as Aberdeen, where no other good kind of rock is to be had. In England the Magnesian limestone is extensively quarried for building purposes. It is of very various qualities, sometimes exceedingly durable, resisting the effects of time and weather, and in other cases decomposing with considerable rapidity. In districts where it occurs, there are churches, and castles, such as Conisbro', built of it, wherein the edges of the stones are as sharp as if fresh from the mason's hands. You cán see the very chisel marks of the men who built the castle, in days soon after the time of William the Conqueror. The Carboniferous Limestone also is an exceedingly durable stone. The Menai bridges were built of it. In Caernarvon castle the preservation of this limestone is well shown. The castle is built of layers of limestone and sandstone, the sandstone having been chiefly derived from the millstone grit, and the limestone from quarries in Anglesey, and on the shores of the Menai Straits. The limestone has best stood the weather. · Sandstone, though durable, is rarely so good as certain limestones, which being somewhat crystalline, and formed to a great extent of Encrinites, also essentially crystalline in structure, have withstood the effect of time.

I have now attempted to give you an idea of the general physical geography of our country, as dependent on its geology. I first described to you the classification of rocks. I divided them into two classes, and one sub-class; consisting of aqueous rocks formed by the action of water, igneous rocks by the action of heat; and of metamorphic rocks which were originally stratified, but have since been acted on by heat and other influences. I then showed you the distribution of these rocks over our country. They have been affected by disturbances and denudations—so that where most disturbed, and hardened, and denuded, there we have mountainous districts; for the greater prominence and ruggedness of surface of these regions arises partly from the hardness of the rocks, partly from the extreme denudation which they have undergone. The Secondary and Tertiary rocks not being so much disturbed, and being younger, have never been so much denuded, and therefore form plains and table-lands. Moreover we saw that over all these surfaces, in addition to the vast amount of erosion which must have been effected in Palæozoic, Secondary and older Tertiary times, renewed denudations accompanied by great cold occurred at a late Tertiary epoch. The result of the last abrasion has been to cover the surface more or less with loose superficial detritus, upon which part of the fertility of portions of the country, and the peculiarity of some of its soils depend. We then passed on to notice what I considered to be a very remarkable result of this last great denudation brought about under the influence of ice, by which the chief part (I by no means say all)—but by which the chief part of the lakes of our country have been formed; and not of our country alone, but of a large part of the northern hemisphere. It is a remarkable thing, indeed, to consider, if true—and I firmly believe it to be true,—that most of those great hollows in which our lakes lie, have been scooped out by the slow and long-continued passage of great sheets of glacier ice, quite comparable to those vast masses that cover the extreme northern and southern regions of the world at this day. The water drainage of the country is likewise seen to be dependent on geological structure. Our large rivers chiefly drain to the east, the smaller ones to the west,—because the great axes of disturbance happened to lie nearer our western than our eastern coasts. Again, the quality of water in these rivers depends, as we have seen, on the nature of the rocks through which they flow, and of the springs by which they are supplied.

Then, when we come to consider the nature of the population inhabiting our island, we find it also to be greatly influenced by this old geology. The earlier tribes have been driven into the more barren mountain regions in the north and west, and so remain to this day—still speaking original languages, but gradually melting up with the great masses of mixed races that came in with later waves of conquest from other parts of Europe. These later races settling down in the more fertile parts of the country, first destroyed and then again began to develop its agricultural resources. In later times they have applied themselves with wonderful energy to turn to use the vast stores of mineral wealth which lie in the central districts. Hence have arisen those densely peopled towns and villages where the manufactures of the country are carried on. Yet in the west, too—in Devon and Cornwall, and in Wales, where some of the great metalliferous and the slaty regions lie—there are busy centres of population, where the mineral products are worked by the original Celtic inhabitants.

It is interesting to go back a little and enquire what
may have been the condition of our country when man
first set foot upon its surface. We know that these
islands of ours have been frequently united to the Continent, and as frequently disunited, partly by elevations
and depressions of the land, and to a great extent, also,
, by denudations. When the earliest human population
came, Britain was probably united to the Continent by
great plains of boulder-drift. Such is the deliberate
opinion of some of our best geologists, and also that
these prehistoric men inhabited our country along with
the great hairy Mammoth, the Rhinoceros, the Cave
Bear, the Lion, and the Hippopotamus,—and perhaps
travelled westwards from the Continent of Europe, along
with these extinct mammalia. But in later times, denudations and alterations of level having again taken
place, our island became again disunited from the
mainland. And now, with all its numerous firths and
inlets, its great extent of coast, its admirable harbours,
our country lies within the direct influence of that
Gulf stream which softens the whole climate of the
west of Europe, and we, a people of mixed race, Celt,
Scandinavian, Saxon, Norman, more or less intermingled in blood, are so happily placed that, in a great
measure, we have the command of the commerce of
the world, and send out fleets of merchandise from
every port. And we are happy, in my opinion, above

all things in this, that.by an old denudation we have been dissevered from the Continent of Europe; for thus it happens that, free from the immediate contact of hostile countries, and almost unbiassed by the influence of peoples of foreign blood, during the long course of years in which our country has never seen the foot of an invader,* we have been enabled so to develop our own ideas of right and wrong, of religion, of political freedom, and of political morality, that we now stand one of the freest countries on the face of the globe, enjoying our privileges under the strongest and freest government in the living world.

* The miserable French descents in Pembrokeshire and Ireland do not deserve the name of invasions.

THE END.

LONDON
PRINTED BY SPOTTISWOODE AND CO.
NEW-STREET SQUARE